Evaluation of U.S. Air Force Preacquisition Technology Development

Committee on Evaluation of U.S. Air Force Preacquisition Technology Development

Air Force Studies Board

Division on Engineering and Physical Sciences

NATIONAL RESEARCH COUNCIL
OF THE NATIONAL ACADEMIES

THE NATIONAL ACADEMIES PRESS
Washington, D.C.
www.nap.edu

THE NATIONAL ACADEMIES PRESS 500 Fifth Street, N.W. Washington, DC 20001

NOTICE: The project that is the subject of this report was approved by the Governing Board of the National Research Council, whose members are drawn from the councils of the National Academy of Sciences, the National Academy of Engineering, and the Institute of Medicine. The members of the committee responsible for the report were chosen for their special competences and with regard for appropriate balance.

This is a report of work supported by Grant FA9550-09-1-0653 between the U.S. Air Force and the National Academy of Sciences. Any opinions, findings, conclusions, or recommendations expressed in this publication are those of the author(s) and do not necessarily reflect the view of the organizations or agencies that provided support for the project.

International Standard Book Number-13: 978-0-309-16275-3
International Standard Book Number-10: 0-309-16275-0

Limited copies of this report are available from:

Air Force Studies Board
National Research Council
500 Fifth Street, N.W.
Washington, DC 20001
(202) 334-3111

Additional copies are available from:

The National Academies Press
500 Fifth Street, N.W.
Lockbox 285
Washington, DC 20055
(800) 624-6242 or (202) 334-3313
(in the Washington metropolitan area)
Internet, http://www.nap.edu

Printed in the United States of America

THE NATIONAL ACADEMIES

Advisers to the Nation on Science, Engineering, and Medicine

The **National Academy of Sciences** is a private, nonprofit, self-perpetuating society of distinguished scholars engaged in scientific and engineering research, dedicated to the furtherance of science and technology and to their use for the general welfare. Upon the authority of the charter granted to it by the Congress in 1863, the Academy has a mandate that requires it to advise the federal government on scientific and technical matters. Dr. Ralph J. Cicerone is president of the National Academy of Sciences.

The **National Academy of Engineering** was established in 1964, under the charter of the National Academy of Sciences, as a parallel organization of outstanding engineers. It is autonomous in its administration and in the selection of its members, sharing with the National Academy of Sciences the responsibility for advising the federal government. The National Academy of Engineering also sponsors engineering programs aimed at meeting national needs, encourages education and research, and recognizes the superior achievements of engineers. Dr. Charles M. Vest is president of the National Academy of Engineering.

The **Institute of Medicine** was established in 1970 by the National Academy of Sciences to secure the services of eminent members of appropriate professions in the examination of policy matters pertaining to the health of the public. The Institute acts under the responsibility given to the National Academy of Sciences by its congressional charter to be an adviser to the federal government and, upon its own initiative, to identify issues of medical care, research, and education. Dr. Harvey V. Fineberg is president of the Institute of Medicine.

The **National Research Council** was organized by the National Academy of Sciences in 1916 to associate the broad community of science and technology with the Academy's purposes of furthering knowledge and advising the federal government. Functioning in accordance with general policies determined by the Academy, the Council has become the principal operating agency of both the National Academy of Sciences and the National Academy of Engineering in providing services to the government, the public, and the scientific and engineering communities. The Council is administered jointly by both Academies and the Institute of Medicine. Dr. Ralph J. Cicerone and Dr. Charles M. Vest are chair and vice chair, respectively, of the National Research Council.

www.national-academies.org

COMMITTEE ON EVALUATION OF U.S. AIR FORCE PREACQUISITION TECHNOLOGY DEVELOPMENT

RICHARD V. REYNOLDS, The VanFleet Group, LLC, *Chair*
DONALD C. FRASER, Charles Stark Draper Laboratory (retired), *Vice Chair*
CHARLES E. ADOLPH, Albuquerque, New Mexico
BRIAN A. ARNOLD, Raytheon Company
FRANCIS J. BAKER, Wright State University
THOMAS W. BLAKELY, Lockheed Martin Aeronautics Company
CLAUDE M. BOLTON, Defense Acquisition University
THOMAS J. BURNS, SET Corporation
LLEWELLYN S. DOUGHERTY, Raytheon Company
RICHARD B.H. LEWIS, Lockheed Martin Corporation
ELLEN M. LORD, AAI Corporation
CHRISTOPHER E. MANUEL, Sierra Nevada Corporation
MATT L. MLEZIVA, Wildwood Strategic Concepts
RONALD E. MUTZELBURG, Alexandria, Virginia
RICHARD L. RUMPF, Rumpf Associates International, Inc.

Staff

CARTER W. FORD, Program Officer and Study Director
SARAH M. CAPOTE, Research Associate
ZEIDA PATMON, Program Associate

AIR FORCE STUDIES BOARD

Preface

The enactment of the Weapon Systems Acquisition Reform Act of 2009 (Public Law 111-23) and the recent revision of Department of Defense Instruction 5000.02 have served to highlight the complexity of the Department of Defense acquisition process.[1] This report serves as a follow-on study to the 2008 National Research Council (NRC) report *Pre-Milestone A and Early-Phase Systems Engineering: A Retrospective Review and Benefits for Future Air Force Systems Acquisition*, which emphasized the role of systems engineering in the acquisition life cycle.[2] This complementary report focuses specifically on the role of maturing technologies and inserting them at the appropriate time in the acquisition cycle.

Leaders in the Air Force responsible for science and technology and acquisition are trying to determine the optimal way to utilize existing policies, processes, and resources to properly document and execute pre-program of record technology development efforts, including opportunities to facilitate the rapid acquisition of revolutionary capabilities and the more deliberate acquisition of evolutionary capabilities.

The Committee on Evaluation of U.S. Air Force Preacquisition Technology Development (see Appendix A for biographical sketches), appointed by the NRC to conduct this study, acknowledges and appreciates the contribution of the mem-

[1] For additional information, see http://www.dtic.mil/whs/directives/corres/pdf/500002p.pdf. Accessed January 14, 2011.

[2] NRC. 2008. *Pre-Milestone A and Early-Phase Systems Engineering: A Retrospective Review and Benefits for Future Air Force Systems Acquisition.* Washington, D.C.: The National Academies Press.

bers of the Air Force Studies Board (AFSB) of the NRC in developing the study's statement of task (see Box 1-1 in Chapter 1) in concert with the Air Force sponsor.

The AFSB was established in 1996 as a unit of the NRC at the request of the United States Air Force. The AFSB brings to bear broad military, industrial, and academic scientific, engineering, and management expertise on Air Force technical challenges and other issues of importance to senior Air Force leaders. The board discusses potential studies of interest, develops and frames study tasks, ensures proper project planning, suggests potential committee members and reviewers for reports produced by fully independent ad hoc study committees, and convenes meetings to examine strategic issues. The board members were not asked to endorse the committee's conclusions or recommendations, nor did they review the final draft of this report before its release, although board members with appropriate expertise may be nominated to serve as formal members of study committees or as report reviewers.

The committee thanks the many people who provided information to the committee, including the guest speakers shown in Appendix B, their organizations, and supporting staff members; and others, including the study sponsor Dr. Steven Walker, Deputy Assistant Secretary of the Air Force for Science, Technology, and Engineering, Office of the Assistant Secretary of the Air Force for Acquisition, and his staff members. The committee is also grateful to the NRC staff members who provided their dedicated support throughout the study.

Richard V. Reynolds, *Chair*
Donald C. Fraser, *Vice Chair*
Committee on Evaluation of U.S. Air Force Preacquisition Technology Development

Acknowledgment of Reviewers

This report has been reviewed in draft form by individuals chosen for their diverse perspectives and technical expertise, in accordance with procedures approved by the National Research Council's Report Review Committee. The purpose of this independent review is to provide candid and critical comments that will assist the institution in making its published report as sound as possible and to ensure that the report meets institutional standards for objectivity, evidence, and responsiveness to the study charge. The review comments and draft manuscript remain confidential to protect the integrity of the deliberative process. We wish to thank the following individuals for their review of this report:

Vinton G. Cerf, Google, Inc.,
Pamela A. Drew, TASC, Inc.,
Charles B. Duke, Xerox Corporation (retired),
Annette J. Krygiel, Great Falls, Virginia,
Paul E. Nielsen, Maj Gen, USAF (retired), Carnegie Mellon University,
Robert E. Schafrik, GE Aircraft Engines, and
Larry D. Welch, Gen, USAF (retired), Institute for Defense Analyses.

Although the reviewers listed above provided many constructive comments and suggestions, they were not asked to endorse the conclusions or recommendations, nor did they see the final draft of the report before its release. The review of this report was overseen by Lawrence T. Papay, PQR, LLC, and Lawrence J. Delaney, Titan Corporation (retired). Appointed by the National Research Council, they were responsible for making certain that an independent examination of this report was carried out in accordance with institutional procedures and that all review comments were carefully considered. Responsibility for the final content of this report rests entirely with the authoring committee and the institution.

Contents

APPENDIXES

Acronyms

A&AS	advisory and assistance services
ACC	adaptive cruise control
ACD&P	Advanced Component Development and Prototypes
AEO	Adaptive Execution Office
AFI	Air Force Instruction
AFMC	Air Force Materiel Command
AFRL	Air Force Research Laboratory
AFSB	Air Force Studies Board
AFSC	Air Force Systems Command
AFSPC	Air Force Space Command
AIP	Acquisition Improvement Plan
AoA	Analysis of Alternatives
ARDEC	Armament Research, Development, and Engineering Center
ATC	Applied Technology Council
ATO	Army Technology Office
CDRT	Capabilities Development for Rapid Transition
CJCS (J8)	Chairman of the Joint Chiefs of Staff (Force Structure, Resources, and Assessment Directorate)
COCOM	(United States) Combatant Command
CP	competitive prototyping
CPM	Critical Path Method
CSAF	Chief of Staff of the Air Force

C/SCSC	Cost/Schedule Control System Criteria
CTE	critical technology element
DAG	*Defense Acquisition Guidebook*
DAPA	Defense Acquisition Performance Assessment
DARPA	Defense Advanced Research Projects Agency
DDR&E	Director, Defense Research and Engineering
DMEA	Defense Microelectronics Activity
DoD	Department of Defense
DP	Development Planning
EFP	explosively formed penetrator
EMD	Engineering and Manufacturing Development (phase)
EPP	"Enhanced" Planning Process
EUREKA	European Research Cooperation Agency
EVM	Earned Value Management
FAR	Federal Acquisition Regulation
FFRDC	Federally Funded Research and Development Center
FNC	Future Naval Capabilities
GAO	Government Accountability Office
GPS	Global Positioning System
ICD	Initial Capabilities Document
IED	improvised explosive device
IPT	Integrated Product Team
IR&D	independent research and development
ISET	Industry System Engineering Team
JCA	Joint Capability Area
JCAAMP	Joint IED Defeat Capability Approval and Acquisition Management Process
JCIDS	Joint Capabilities Integration and Development System
JGRE	Joint Ground Robotics Enterprise
JHU/APL	Johns Hopkins University/Applied Physics Laboratory
JIEDD	Joint Improvised Explosive Device Defeat
JIEDDO	Joint Improvised Explosive Device Defeat Organization
JIPT	JIEDDO Integrated Process Team
JPG	Joint Programming Guidance
JR2AB	JIEDDO Requirements, Resources, and Acquisition Board

JRAC	Joint Rapid Acquisition Cell
JROC	Joint Requirements Oversight Council
JSF	Joint Strike Fighter
JUONS	Joint Urgent Operational Needs Statement
LCC	life-cycle cost
MAJCOM	Major Command
MDA	Milestone Decision Authority
MDAP	Major Defense Acquisition Program
MDD	Material Development Decision
MRL	Manufacturing Readiness Level
NASA	National Aeronautics and Space Administration
NDAA	National Defense Authorization Act
NLOS-LS	Non-Line-of-Sight Launch System
NRC	National Research Council
NRL	Naval Research Laboratory
NSAC	National Small Arms Center
O&S	operations and support
OMB	Office of Management and Budget
ONR	Office of Naval Research
OSD	Office of the Secretary of Defense
OTA	Other Transactions Agreement
OUSD (AT&L)	Office of the Under Secretary of Defense (Acquisition, Technology, and Logistics)
OUSD (C)	Office of the Under Secretary of Defense (Comptroller)
PDR	Preliminary Design Review
PE	Program Element
PERT	Program Evaluation Review Technique
POM	Program Objective Memorandum
PPBE	Planning, Programming, Budgeting, and Execution
PPBES	Planning, Programming, Budgeting, and Execution System
R&D	research and development
RAM	Requirements Analysis and Maturation
RDT&E	Research, Development, Test, and Evaluation
ROVER	Remotely Operated Video Enhanced Receiver
RRTO	Rapid Reaction Technology Office

S&T	science and technology
SAF	Secretary of the Air Force
SAF/AQ	Secretary of the Air Force (Acquisition)
SAF/AQX	Secretary of the Air Force (Acquisition Integration)
SMC	Space and Missile Systems Center
SNC	Sierra Nevada Corporation
SOTVS	Special Operations Tactical Video System
SPG	Strategic Planning Guidance
SPO	System Program Office
SRR	System Requirements Review
TARDEC	Tank Automotive Research, Development, and Engineering Center
TDS	Technology Development Strategy
TIG	technology interest group
TRA	Technology Readiness Assessment
TRADOC	Training and Doctrine Command
TRL	Technology Readiness Level
TSPR	Total System Performance Responsibility
USAF	United States Air Force
USN	United States Navy
VADER	Vehicle and Dismount Exploitation Radar
WSARA	Weapon Systems Acquisition Reform Act
XR	Product Centers' Development Planning Organization

Summary

From the days of biplanes and open cockpits, the air forces of the United States have relied on the mastery of technology to ensure what, in 1921, Giulio Douhet called "the command of the air."[1] And while the weapons of air warfare have changed, the vital importance of technological superiority to the United States Air Force has not.

Although evidence exists—for example, Government Accountability Office (GAO)[2] reports, failed programs, and programmatic breaches in cost, schedule, and technical performance—that the Air Force is currently struggling to incorporate technology in its major systems acquisitions successfully, it is important to note that the path toward technological supremacy has never been a smooth one.[3,4,5] Describing the technological travails that he faced 75 years ago while building the

[1]Giulio Douhet. 1921. *The Command of the Air.* Coward-McCann, 1921, translated 1942 by Dino Ferrari. Washington, D.C.: Printing-Office for the Use of the War Department.

[2]Effective July 7, 2004, the legal name of the General Accounting Office was changed to "Government Accountability Office." The change, which better reflects the modern professional services organization that the GAO has become, is a provision of the GAO Human Capital Reform Act of 2004, Pub. L. 108-271, 118 Stat. 811 (2004).

[3]GAO. 1999. *Best Practices: Better Management of Technology Development Can Improve Weapon System Outcomes.* GAO/NSIAD-99-162. Washington, D.C.: GAO.

[4]GAO. 2001. *Best Practices: Better Matching of Needs and Resources Will Lead to Better Weapon System Outcomes.* GAO-01-288. Washington, D.C.: GAO.

[5]GAO. 2006. *Best Practices: Stronger Practices Needed to Improve DOD Technology Transition Processes.* GAO-06-883. Washington, D.C.: GAO.

early Army Air Corps, General of the Air Force Henry H. ("Hap") Arnold told of a reality not unlike that of today:

> Planes became obsolescent as they were being built. It sometimes took five years to evolve a new combat airplane, and meanwhile a vacuum could not be afforded. . . . I also had trouble convincing people of the time it took to get the "bugs" out of all the airplanes. Between the time they were designed and the time they could be flown away from the factory stretched several years. For example . . . the B-17 was designed in 1934, but it was 1936 before the first test article was delivered. The first production article was not received by the Air Corps until 1939. You can't build an Air Force overnight.[6]

Yet the 5 years from design to operation that General Arnold described have now stretched in some cases to 20 years and more, and cost has increased similarly. Much of the delay and cost growth afflicting modern Air Force programs is rooted in the same area that plagued General Arnold: the incorporation of advanced technology into major systems acquisition.

STUDY APPROACH

In response to a request from the Deputy Assistant Secretary of the Air Force for Science, Technology, and Engineering, the National Research Council (NRC) formed the Committee on Evaluation of U.S. Air Force Preacquisition Technology Development.[7] The statement of task for this study is as follows:

- Examine appropriate current or historical DoD [Department of Defense] policies and processes, including the PPBES [Planning, Programming, Budgeting, and Execution System], DoD Instruction 5000.02, the Air Force Acquisition Improvement Plan, JCIDS [Joint Capabilities Integration and Development System], and DoD and Air Force competitive prototyping policies to comprehend their impact on the execution of pre-program of record technology development efforts.
- Propose any changes to the Air Force workforce, organization, policies, processes and resources, if any, to better perform preacquisition technology development. Specific issues to consider include:
  —Resourcing alternatives for Pre-Milestone B activities
  —The role of technology demonstrations

[6]H.H. Arnold. 1949. *Global Mission.* New York, N.Y.: Harper, pp. 178-179, 193.

[7]For purposes of this study, preacquisition technology development involves: (1) determining what advanced technology is needed, by when, and for what purpose; (2) assessing and balancing the feasibility, benefits, and costs of developing, inserting, and deploying the technology; (3) deciding the best path to reduce the risk in achieving the desired results within estimated cost and schedule; and (4) responsibly executing technology maturation activities.

- Study and report on industry/Government best practices to address both evolutionary (deliberate) and revolutionary (rapid) technology development.
- Identify potential legislative initiatives, if any, to improve technology development and transition into operational use.

With the task in mind, the committee began a process of evidence gathering in which efforts were focused on gaining a current and accurate picture of the situation in the air, space, and cyberspace domains, through documentary research and through interactions with a large number of government agencies and offices. The committee conducted four data-gathering meetings at which input to the study was provided by the following: senior Air Force leaders, including representatives of several Air Force Major Commands; representatives from the other military departments; senior officials in the Office of the Secretary of Defense (OSD); professional staff members from key congressional oversight committees; and senior industry executives. This effort was followed by an exploration of best practices in which the lessons of technological success stories from academia, government, and industry were studied.

Early in the study, the committee developed a framework, the "Three Rs," for organizing its findings and recommendations. The framework describes characteristics that, in the committee's judgment, need to be addressed fully in order for successful technology development to occur. That framework is composed of the following:

1. *Requirements*—clear, realistic, stable, trade-off tolerant, and universally understood;
2. *Resources*—adequate and stable, and including robust processes, policies, and budgets; and
3. *The Right People*—skilled, experienced, and in sufficient numbers, with stable leadership.

On the basis of this framework, the committee developed a number of findings and recommendations that are presented in Chapters 2 through 4; the full set of recommendations is provided below.[8] In keeping with its statement of task, the committee studied the current state of Air Force technology development and the environment in which technology is acquired, and then it looked at best practices from both government and industry. Because the resulting recommendations are in all cases within the power of the Air Force to implement, the committee chose

[8]The findings and recommendations retain their original numbering regardless of where they appear in the text: for example, Recommendation 4-1 is the first recommendation in Chapter 4.

not to specify any near-term legislative initiatives, the possibility of which was envisioned in the statement of task.

RECOMMENDATIONS

Requirements

There is very little new in the management of technology development. Important lessons have been learned before by the Air Force, and, regrettably, many seem to have been forgotten. At the same time, industry has learned—and the Air Force is seemingly having to relearn—that simultaneously developing new technology within an acquisition program is a recipe for disaster. Just as in Hap Arnold's day, requirements that are unclear, unrealistic, or unstable inhibit successful technology insertion. In 1922, General Arnold studied the biplane pursuit craft that represented the technology of the day, and he came to understand fully the dangers of shifting requirements in an acquisition program:

> [O]nce production had begun, the line must be allowed to run undisturbed. Any new improvements should wait until a specified point. . . . Mass production requires certain sacrifices in technological advancement, [Arnold] reported; the trick was to be aware of what was needed before production began, "and then to stick to it for a certain period even though it can be improved, until such time as the improvement can be incorporated without materially affecting production."[9]

Although the dangers posed by shifting requirements are well known, the temptation to improve systems in development can be hard to resist. This temptation only increases as product development cycles lengthen from years to decades.[10]

RECOMMENDATION 4-1
To ensure that technologies and operational requirements are well matched, the Air Force should create an environment that allows stakeholders—warfighters, laboratories, acquisition centers, and industry—to trade off technologies with operational requirements prior to Milestone B.

RECOMMENDATION 4-2
To enable (1) a more disciplined decision-making process and (2) a forum in

[9]Dik Alan Daso. 2001. *Hap Arnold and the Evolution of American Airpower.* Washington, D.C.: Smithsonian Books, p. 98.

[10]The recommendations in this report apply to each of the three operational domains of the Air Force: air, space, and cyberspace. At the same time, each domain is unique due to its particular characteristics and the unique environments in which it operates.

which all stakeholders—those from the science and technology (S&T), acquisition, and warfighting Major Command (MAJCOM) communities—can focus their attention jointly on critical technology development questions and then make tough strategy and resource calls efficiently at a level where the decisions are most likely to stick, the Air Force should consider adopting a structure similar to the Navy's S&T Corporate Board and Technology Oversight Group and the Army Technology Objectives Process and Army S&T Advisory Group. A committee-developed notional organization for Air Force consideration (Figure S-1) addresses this potential and is tailored to Air Force missions and organization. In addition, the Air Force should consider allocating funding for technology development, including funding for 6.4, or advanced component development and prototypes, to the Air Force Materiel Command (AFMC) and Air Force Space Command (AFSPC), unless precluded by law from doing so.

In the opinion of the committee, this recommendation to add another organization to the Headquarters Air Force does not diminish the statutory and mission responsibilities of the Assistant Secretary of the Air Force (Acquisition) (SAF/AQ)[11] and is justified by the seriousness of the need. In the committee's judgment, no other approach would meet the need to bring together the S&T, acquisition, and warfighting MAJCOM communities at a level that could make the difficult decisions. The fundamental premise of Recommendation 4-2 is the importance of technology to the Air Force, as described in the introductory paragraphs of Chapter 1 and reiterated in introductory statements in Chapter 2. Findings 2-8 and 2-9 (first presented in Chapter 2 and then repeated in the context of the associated recommendation in Chapter 4) identify significant shortfalls in decision making for Air Force technology development and transition—that is, the lack of a process for technology transition and, at a higher level, the lack of a service-wide unifying S&T strategy to guide investments—which, in the judgment of the committee, need to be addressed. The structure proposed in Recommendation 4-2 would give SAF/AQ greater leverage to ensure that the right technology is being developed, matured, and transitioned. Furthermore, the cross-domain character of technology development, addressed in Chapters 1 through 3 of this report, presents challenges that the recommended S&T Board could address efficiently with a diverse set of stakeholders at the table. Finally, given the ever-increasing complexity and budget implications of new weapon systems, in the opinion of the committee the status quo is not acceptable.

[11]The SAF/AQ's responsibilities are specified under the Goldwater-Nichols Department of Defense Reorganization Act of 1986 (Public Law 99-433) and Headquarters Air Force Mission Directive 1-10, April 8, 2009.

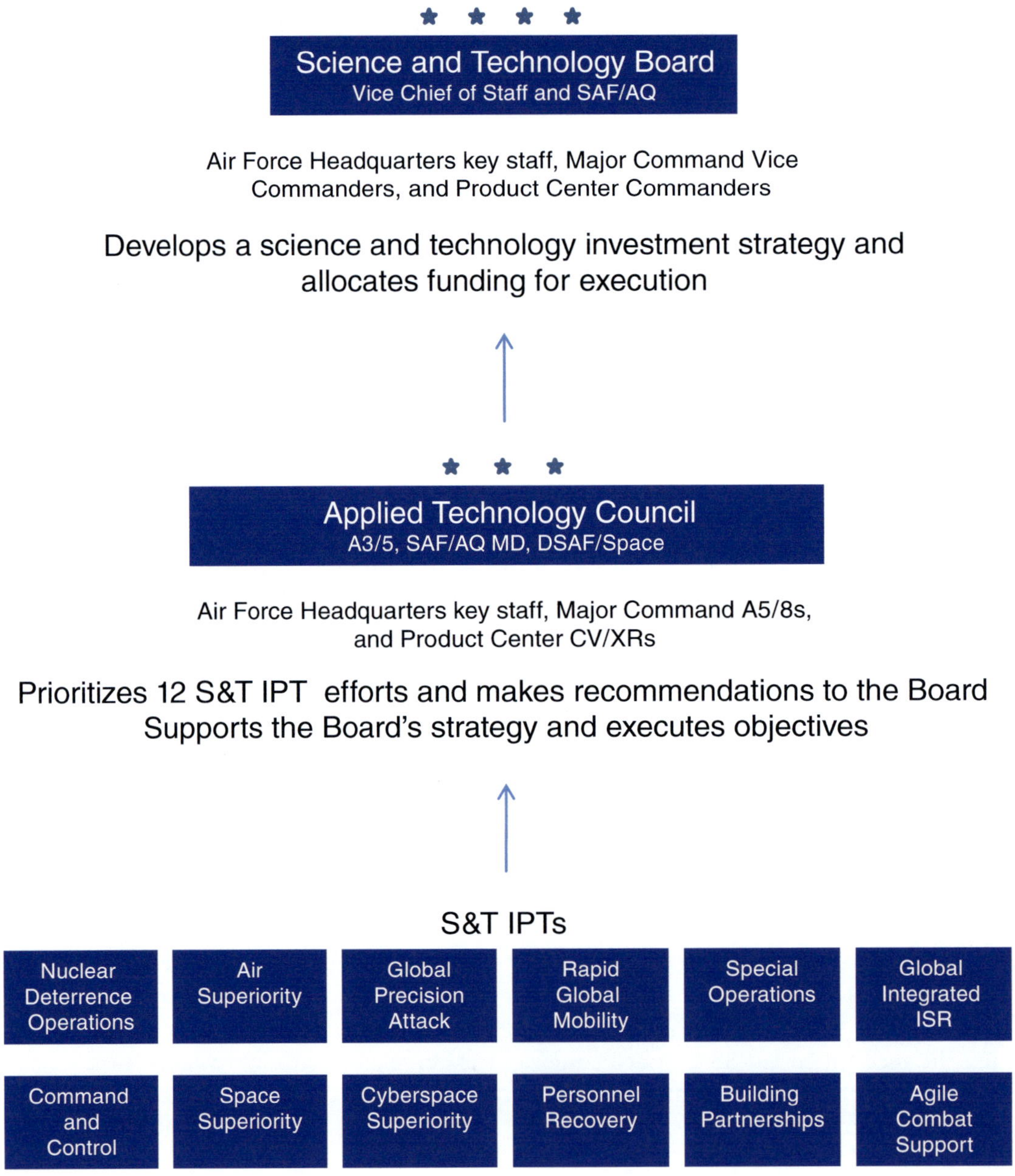

FIGURE S-1
Notional science and technology governance.

Resources

None of the many Air Force presenters who briefed the committee was able to articulate an Air Force-level, integrated S&T strategy, nor could any identify a single office with authority, resources, and responsibility for all S&T initiatives across the Air Force. Instead, there appears to be an assortment of technology "sandboxes," in which various players work to maximize their organizational self-interest, as they perceive it. In such a system, optimization will always take place at the subunit level, with less regard for the health of the overarching organization.

Processes and procedures to facilitate the successful integration of technology into major system acquisitions were developed by the Air Force long ago, and some were in existence within relatively recent memory. But for various reasons, many of these were ended or allowed to atrophy. Chief among these were initiatives such as the historical Air Force Systems Command's Vanguard process and the acquisition Product Centers' Development Planning Organizations (XRs), which for decades formed a crucial link between warfighter requirements on the one hand and laboratory and industry capabilities on the other. Funding for Development Planning was zeroed out a decade ago, and the negative impacts of that decision are now clear. Other such activities, like systems engineering and Applied Technology Councils, also declined in importance in some arenas, with similar harmful results.

Other processes that do exist elsewhere need to be adopted more accurately, effectively, and consistently by the Air Force.[12] For example, Technology Readiness Levels (TRLs) are a process created by the National Aeronautics and Space Administration (NASA) in which technology is incorporated in operational environments only after it is proven to be mature. The committee observed many examples from industry, NASA, and the AFSPC in which disciplined and objective adherence to rigorous technology readiness principles led to the successful incorporation of new technology into major systems. The Air Force as a whole, however, has yet to demonstrate full commitment to TRL principles.

RECOMMENDATION 4-3

Since DoD Instruction 5000.02 incorporates increased pre-Milestone B work, the Air Force should bring Pre-Milestone B work content back into balance with available resources by some combination of (1) DoD Instruction 5000.02 tailoring and/or (2) additional expertise, schedule, and financial resources. Examples of expanded content include competitive prototyping, demonstrating

---

[12]DoD. 2010. *Quadrennial Defense Review Report.* Washington, D.C.: Department of Defense. Available at http://www.defense.gov/qdr/images/QDR_as_of_12Feb10_1000.pdf. Accessed August 12, 2010.

technology in operationally relevant environments, and completing preliminary design prior to Milestone B.

RECOMMENDATION 4-4
Knowledgeable, experienced, and independent technical acquisition professionals outside the program office should conduct technology, manufacturing, and integration assessments using consistent, rigorous, and analytically based standards. While the Weapon Systems Acquisition Reform Act of 2009 (Public Law 111-23) requires this effort to be executed at the OSD level, this organic capability needs to be developed and assigned to the AFMC and the AFSPC. Once this capability has been effectively demonstrated by the Air Force, legislative relief should be sought.

RECOMMENDATION 4-5
To increase the likelihood of acquisition success, the Air Force should enter Engineering and Manufacturing Development (Milestone B) only with mature technologies—that is, with technologies at TRL 6 or greater.

RECOMMENDATION 4-6
The Air Force should drive greater collaboration between warfighters (to include joint and coalition partners), laboratories, developers, and industry. One approach is to establish collaboration forums similar to the Ground Robotics Consortium and the Army Armament Research, Development, and Engineering Center's National Small Arms Center.

The Right People

The literal decimation of the Air Force acquisition workforce over the past two decades is well known. Although a workforce can be slashed in a very short time, rebuilding it in terms of knowledge, skills, and experience can take decades. The Air Force seems to have recognized the damage done in this regard and is moving to reverse course, with substantial hiring of acquisition specialists at both the trainee and the journeyman level.

Importantly, the present study is focused not on acquisition broadly, but rather on the specific intersection between technology and major systems acquisition. With this in mind, the recommendation in the area of the "Third R" is focused on continuing the reinvigoration of Development Planning. The Air Force has recognized the tremendous cost imposed by the elimination of the XRs, and it has moved to begin to recoup its losses by restoring funding and people to the vital Development Planning function. This is a good start, but as with rebuilding the general acquisition force, more needs to be done, and progress will occur slowly.

RECOMMENDATION 4-7

The Air Force should accelerate the re-establishment of the Development Planning organizations and workforce and should endow them with sufficient funds, expertise, and authority to restore trust in their ability to lead and manage the technology transition mission successfully.

A FINAL COMMENT: "THE DEATH SPIRAL," AND THE CASE OF THE 137-PERSON REVIEW TEAM

Over the months of this study, the committee found substantial evidence of a condition that, although perhaps beyond the strict limits of the statement of task, was tightly interwoven with the issues of technology and major systems acquisition. That condition is the pervasive lack of trust apparent in the entire DoD systems acquisition process. This lack of trust is both cause and effect, in some ways being created by ineffective technology insertion and in other ways creating its own inefficiencies throughout the process.

Exactly where the Death Spiral began is open to debate, a chicken-or-egg type of argument. Did technological failure and acquisition disappointments create the massive growth of oversight at every level, which slows the acquisition process and saps its energy? Or is it equally plausible that the growth of oversight in fact *creates* the very failures in cost, schedule, and performance that it is designed to prevent?

What *is* certain is that an unhealthy and self-perpetuating spiral involving the loss of trust and the growth of oversight does in fact exist. One presenter to the committee spoke of a program to which the contractor had assigned 80 engineers, who stood stunned as a government review team arrived with 137 participants, most of them junior military and civilian employees.[13] As was described in the 2008 NRC report *Pre-Milestone A and Early-Phase Systems Engineering: A Retrospective Review and Benefits for Future Air Force Systems Acquisition*:

> The DoD management model is based on a lack of trust. Quantity of oversight has replaced quality. There is no clear line of responsibility, authority, or accountability. Oversight is preferred to accountability. . . . The complexity of the acquisition process increases cost and draws out the schedule.[14]

[13]Neil Kacena, Vice President, Advanced Development Programs Deputy, Lockheed Martin Aeronautics Company. 2010. "Technology Development: Approaches and Challenges." Presentation to the committee, May 12, 2010.

[14]NRC. 2008. *Pre-Milestone A and Early-Phase Systems Engineering: A Retrospective Review and Benefits for Future Air Force Systems Acquisition.* Washington, D.C.: The National Academies Press. Footnote, p. 13.

The interaction of technology and acquisition management described in the remainder of this report is a complex subject, and as such it is resistant to easy fixes. Nevertheless, beginning in some way to rebuild the sense of trust that was once present among the participants in these processes would seem a logical place to begin.

1

Preacquisition Technology Development for Air Force Weapon Systems

The National Research Council (NRC) issued a report in 2008 entitled *Pre-Milestone A and Early-Phase Systems Engineering: A Retrospective Review and Benefits for Future Air Force Systems Acquisition* (hereinafter referred to as the Kaminski report, after Paul G. Kaminski, the chair of that report's study committee).[1] The Kaminski report emphasized the importance of systems engineering early in the Department of Defense (DoD) acquisition life cycle and urged the revitalization of the systems engineering and Development Planning (DP) disciplines throughout the DoD.[2]

No less important to the future combat capability of the armed services is the development of new, cutting-edge technology. This is particularly true for the United States Air Force (USAF), which from its very inception has sought to capitalize on technological and scientific advances. Even before there was a United States Air Force, the Chief of Staff of the U.S. Army Air Corps Henry H. "Hap" Arnold was committed to giving his forces a decisive technological edge:

> Arnold . . . intended to leave to his beloved air arm a heritage of science and technology so deeply imbued in the institution that the weapons it would fight with would always be the best the state of the art could provide and those on its drawing boards would be prodigies of futuristic thought.[3]

---

[1]NRC. 2008. *Pre-Milestone A and Early-Phase Systems Engineering: A Retrospective Review and Benefits for Future Air Force Systems Acquisition.* Washington, D.C.: The National Academies Press.

[2]Ibid.

[3]Neil Sheehan. 2009. *A Fiery Peace in a Cold War: Bernard Schriever and the Ultimate Weapon.* New York, N.Y.: Random House, p. xvi.

General Arnold succeeded in his dream of building the foundations of an Air Force that was second to none technologically. Dramatic innovations in aeronautics and later in space were fielded, with schedules that today seem impossible to achieve. The first U-2 flew just 18 months after it was ordered in 1953, and it was operational just 9 months after that first flight.[4] The SR-71, even more radical, was developed with similar speed, going from contract award to operational status in less than 3 years.[5] In the space domain, innovation was pursued with similar speed: for example, the Atlas A, America's first intercontinental ballistic missile, required only 30 months from contract award in January 1955 to first launch in June 1957.[6]

At that time, the American military and defense industry set the standard in the effective management of new technology. In fact, the entire field known today as "project management" springs from the management of those missile development programs carried out by the Air Force, the United States Navy, and later the National Aeronautics and Space Administration (NASA). Tools used routinely throughout the project management world today—Program Evaluation Review Technique (PERT) and Critical Path Method (CPM) scheduling systems, Earned Value Management (EVM), Cost/Schedule Control System Criteria (C/SCSC), for example—trace back directly to the work of the Air Force, the Navy, and NASA in those years.[7]

Clearly those days are gone. The Kaminski report cites compelling statistics that describe dramatic cost and schedule overruns in specific, individual programs. Taken all together, the picture for major system acquisition is no better:

> The time required to execute large, government-sponsored systems development programs has more than doubled over the past 30 years, and the cost growth has been at least as great.[8]

STATEMENT OF TASK AND COMMITTEE FORMATION

The Air Force requested that the National Research Council review current conditions and make recommendations on how to regain the technological expertise so characteristic of the Air Force's earlier years. Such outside studies have long been part of the Air Force's quest for improvement in technology. For example,

---

[4]Clarence L. "Kelly" Johnson. 1989. *More Than My Share of It All.* Washington, D.C.: Smithsonian Institution Scholarly Press.

[5]Ibid.

[6]For additional information, see http://www.fas.org/nuke/guide/usa/icbm/sm-65.htm. Accessed May 10, 2010.

[7]For additional information, see http://www.mosaicprojects.com.au/PDF_Papers/P050_Origins_of_Modern_PM.pdf. Accessed May 10, 2010.

[8]NRC. 2008. *Pre-Milestone A and Early-Phase Systems Engineering: A Retrospective Review and Benefits for Future Air Force Systems Acquisition.* Washington, D.C.: The National Academies Press, p. 14.

General Hap Arnold, in his autobiography, described the difficulties that he faced leading the pre-World War II United States Army Air Corps:

> Still, in spite of our smallness and the perpetual discouragements, it was not all bad. Progress in engineering, development, and research was fine. At my old stamping grounds in Dayton, I found the Materiel Division doing an excellent job within the limits of its funds. [General Oscar] Westover was calling on the National Research Council for problems too tough for our Air Corps engineers to handle. . . .[9]

Early in the present study, the Air Force pointed to three elements of the existing acquisition process as examples of things that required improvement. First, evolutionary technology transition has suffered from less-than-adequate early (pre-Milestone A) planning activities that manifest themselves later in problems of cost, schedule, and technical performance. Second, revolutionary transition too often competes with evolutionary transition, as reflected in efforts to rush advanced technology to the field while failing to recognize and repair chronic underfunding of evolutionary Air Force acquisition efforts. Third, there appears to be no single Air Force research and development (R&D) champion designated to address these issues.

Although technology plays a part in all Air Force activities, from operations to sustainment and systems modification, the task for the present study was targeted at the development and acquisition of new major systems. Accordingly, this study focuses on how to improve the ability to specify, develop, test, and insert new technology into new Air Force systems, primarily pre-Milestone B. Box 1-1 contains the statement of task for this study.

To address the statement of task, the Committee on Evaluation of U.S. Air Force Preacquisition Technology Development was formed. Biographical sketches of the committee members are included in Appendix A.

THE PARAMETERS OF THIS STUDY

The statement of task specifically requires assessment of relevant DoD processes and policies, both current and historical, and invites proposed changes to Air Force workforce, organization, policies, processes, and resources. Issues of particular concern include resourcing alternatives for pre-Milestone B activities and the role of technology demonstrations. Previous NRC studies, and studies by other groups,

[9]H.H. Arnold. 1949. *Global Mission.* New York, N.Y.: Harper, p. 165.

have addressed portions of the material covered in this report.[10,11,12,13,14,15,16] Importantly, however, no previous report was expressly limited to addressing the topic of early-phase technology development that is the focus of this study.

Studies of major defense systems acquisition are certainly not in short supply. Over the previous half-century there have been literally scores of such assessments, and their findings are remarkably similar: Weapons systems are too expensive, they take too long to develop, they often fail to live up to expectations. A central question for a reader of this report has to be—What makes *this* study any different?

The answer is, in a word, technology, and its development and integration into Air Force systems. From those early days of Hap Arnold, it was the capable development, planning, and use of technology that set the Air Force and its predecessors apart from the other services. That technological reputation needs to be preserved—some would say recaptured—if the Air Force is to continue to excel in the air, space, and cyberspace domains discussed later in this chapter.

One cause of this technological challenge is that, for a variety of reasons, the Air Force has lost focus on technology development over the past two decades. The Kaminski report makes clear that Air Force capabilities in the critical areas of systems engineering and Development Planning were allowed to atrophy. These declines had their origins in legislative actions, financial pressures, demographics, workforce development, and a host of other sources. But altogether, they led to a

[10]NRC. 2008. *Pre-Milestone A and Early-Phase Systems Engineering: A Retrospective Review and Benefits for Future Air Force Systems Acquisition.* Washington, D.C.: The National Academies Press.

[11]NRC. 2010. *Achieving Effective Acquisition of Information Technology in the Department of Defense.* Washington, D.C.: The National Academies Press.

[12]Assessment Panel of the Defense Acquisition Performance Assessment Project. 2006. *Defense Acquisition Performance Assessment Report.* A Report by the Assessment Panel of the Defense Acquisition Performance Assessment Project for the Deputy Secretary of Defense. Available at https://acc.dau.mil/CommunityBrowser.aspx?id=18554. Accessed June 10, 2010.

[13]Gary E. Christle, Danny M. Davis, and Gene H. Porter. 2009. *CNA Independent Assessment.Air Force Acquisition: Return to Excellence.* Alexandria, Va.: CNA Analysis & Solutions.

[14]Business Executives for National Security. 2009. *Getting to Best: Reforming the Defense Acquisition Enterprise. A Business Imperative for Change from the Task Force on Defense Acquisition Law and Oversight.* Available at http://www.bens.org/mis_support/Reforming%20the%20Defense.pdf. Accessed June 10, 2010.

[15]USAF. 2008. *Analysis of Alternative (AoA) Handbook: A Practical Guide to Analysis of Alternatives.* Kirtland Air Force Base, N.Mex.: Air Force Materiel Command's (AFMC's) Office of Aerospace Studies. Available at http://www.oas.kirtland.af.mil/AoAHandbook/AoA%20Handbook%20Final.pdf. Accessed June 10, 2010.

[16]DoD. 2009. *Technology Readiness Assessment (TRA) Deskbook.* Prepared by the Director, Research Directorate (DRD), Office of the Director, Defense Research and Engineering (DDR&E). Washington, D.C.: Department of Defense. Available at http://www.dod.mil/ddre/doc/DoD_TRA_July_2009_Read_Version.pdf. Accessed September 2, 2010.

BOX 1-1
Statement of Task

The NRC will:

1. Examine appropriate current or historical Department of Defense (DoD) policies and processes, including the Planning, Programming, Budgeting, and Execution System, DoD Instruction 5000.02, the Air Force Acquisition Improvement Plan, the Joint Capabilities Integration and Development System, and DoD and Air Force competitive prototyping policies to comprehend their impact on the execution of pre-program of record technology development efforts.

2. Propose changes to the Air Force workforce, organization, policies, processes and resources, if any, to better perform preacquisition technology development. Specific issues to consider include:

   a. Resourcing alternatives for Pre-Milestone B activities
   b. The role of technology demonstrations

3. Study and report on industry/Government best practices to address both evolutionary (deliberate) and revolutionary (rapid) technology development.

4. Identify potential legislative initiatives, if any, to improve technology development and transition into operational use.

decline in the Air Force's ability to successfully integrate technology into weapons systems in a timely and cost-effective manner.

This study was also commissioned at a time of increased interest in technology management outside the Air Force and beyond the DoD. Disappointed by acquisition programs that underperformed in terms of cost and schedule, Congress enacted the Weapon Systems Acquisition Reform Act of 2009 (WSARA; Public Law 111-23). One of WSARA's major goals is to reduce the likelihood of future programmatic failure by reducing concurrency (i.e., the simultaneity of two or more phases of the DoD acquisition process), thus ensuring that systems and major subsystems are technologically mature before entering production. This congressional intent is apparent in WSARA's emphasis on systems engineering and development planning, and in other mandates' requirements for technology demonstrations, competitive prototypes, and preliminary design reviews earlier in the acquisition cycle, before costly system development and production decisions are made.

COMMITTEE APPROACH TO THE STUDY

Throughout the study, the committee met with numerous Air Force stakeholders to gain a fuller understanding of the sponsor's needs and expectations relating to the elements contained in the statement of task. The full committee met four times to receive briefings from academic, government, and industry experts in technology development, and it conducted a number of visits during which subgroups of the committee met with various stakeholders. The committee met two additional times to discuss the issues and to finalize its report. Appendix B lists specific meetings, individual participants, and participating organizations.

The almost absurd complexity of Figure 1-1 illustrates how daunting the DoD acquisition system is. A clearer image is not available. Given the incredible intricacy of the system, coupled with the relatively short time line of the study, the committee endeavored at the outset to distill this complexity into a basic touchstone mission statement: How to improve the Air Force's ability to specify, develop, test, and insert new technology into Air Force systems.

THREE DOMAINS OF THE AIR FORCE

> The mission of the United States Air Force is to fly, fight, and win . . . in air, space, and cyberspace.[17]

This mission statement, set forth in a joint September 2008 letter from the Secretary of the Air Force and the Air Force Chief of Staff, emphasizes the importance of all three domains in which the Air Force must operate. Each domain involves special considerations and challenges. Consequently, each of the three domains represents a unique environment in terms of science and technology (S&T) and major systems acquisition.

Air

The air domain is perhaps most frequently associated with Air Force major systems acquisition. It is characterized by relatively low (and declining) numbers of new major systems, with a relatively small number of industry contractors competing fiercely to win each new award. In this realm, relationships between government and industry tend to be at arm's length and sometimes adversarial. The duration

[17]"U.S. Air Force Mission Statement and Priorities." September 15, 2008. Available at http://www.af.mil/information/viewpoints/jvp.asp?id=401. Accessed May 19, 2010.

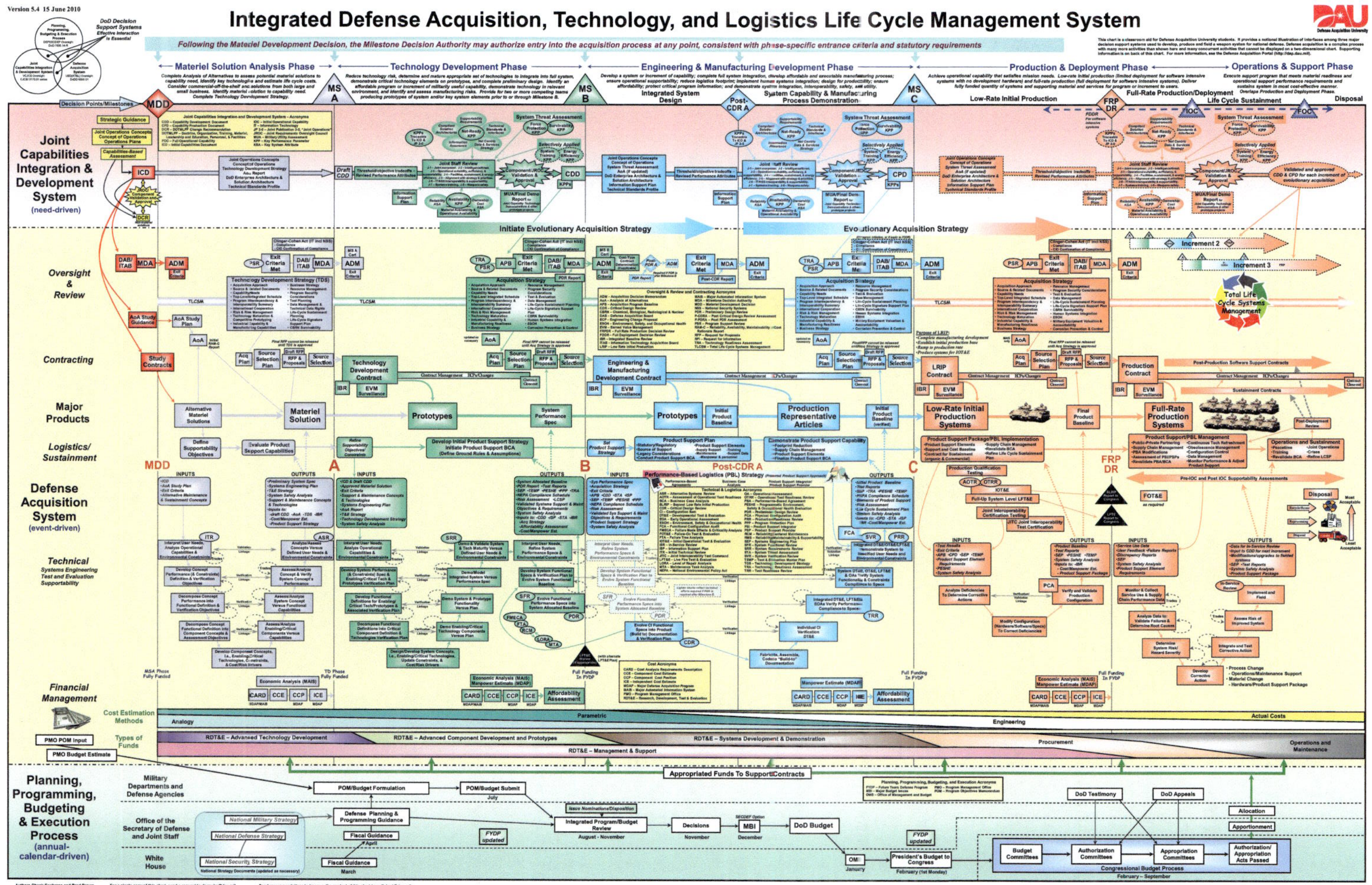

FIGURE 1-1
Department of Defense acquisition process. SOURCE: Defense Acquisition University. Available at https://ilc.dau.mil/default_nf.aspx. Accessed June 11, 2010.

of acquisition programs tends to be long, often measured in decades, whereas "buy quantities" have declined dramatically over time.[18]

In the air domain, not all technology insertion takes place prior to the initial delivery of a system. Aircraft stay in the active inventory for far longer periods than in years past. For example, the *newest* B-52 is now 48 years old, and most KC-135 aerial refueling tankers are even older. The advancing age of such aircraft means that numerous carefully planned and executed technology insertions are therefore required to upgrade and extend the lives of aging fleets.

These post-acquisition technology-based activities are in themselves both militarily necessary and economically significant, and they are increasingly characteristic of the air domain. The financial impacts of these activities are especially noteworthy. For example, the periodic overhauls of B-2 stealth bombers require 1 year and $60 million—per aircraft. Yet the military imperatives leave the Air Force little choice:

> "Although there is nothing else like the B-2, it's still a plane from the 1980s built with 1980s technology," said Peter W. Singer, a senior fellow at the Brookings Institution think tank. Other countries have developed new ways to expose the B-2 on radar screens, so the Air Force has to upgrade the bomber in order to stay ahead. "Technology doesn't stand still; it's always moving forward," he said. "It may cost an arm and a leg, but you don't want the B-2 to fall behind the times."[19]

Major aeronautical systems are thus characterized by large expenditures for research and development, as well as for the initial procurement and periodic updates of the end items themselves. But the largest expenditures tend to be in operations and support (O&S) costs over the life of a system. This is increasingly true, as systems are kept in the inventory for longer periods: The initial purchase price of the Air Force's B-52 bomber was about $6 million in 1962; that sum is dwarfed by the resources required to operate and support the bomber over the last half-century.[20]

Space

"Space is different." This idea was raised repeatedly during the course of this study. The challenges of the space world are, in fact, significantly different from

[18]Following, for example, are approximate buy quantities in the world of multi-engine bombers, from World War II to today: 18,000 B-24s; 12,000 B-17s; 4,000 B-29s; 1,200 B-47s; 800 B-52s; 100 B-1s; 21 B-2s. Similar purchasing patterns exist for fighter and cargo aircraft.

[19]W.J. Hennigan. 2010. "B-2 Stealth Bombers Get Meticulous Makeovers." *Los Angeles Times*, June 10. Available at http://articles.latimes.com/2010/jun/10/business/la-fi-stealth-bomber-20100610. Accessed June 22, 2010.

[20]Information available at https://acc.dau.mil/CommunityBrowser.aspx?id=241468. Accessed May 18, 2010.

those in the air domain. Some of those differences are obvious. For example, space presents an extraordinarily unforgiving environment in which few "do-overs" are possible; by comparison, the air domain offers the luxury of maturing complex technology prototypes through a sequence of relatively rapid "fly-fix-fly" spirals during the development phase. This "spiral development process" for aircraft allows the refinement of complex technologies through responses to empirical observations. In contrast, the space domain offers few such opportunities. Furthermore, in contrast with aircraft production lines, space systems tend to be craft-produced in small quantities by skilled craftspeople.

Additionally, in the space domain there are few if any "flight test vehicles," in that every launch is an operational mission, and failures in the always-critical launch phase tend to be spectacular and irreversible. Therefore, space development programs rely on "proto-qualification" or engineering models and must "test like you fly" in order to maximize the opportunity for on-orbit mission success. Once on orbit, a space platform must work for its lifetime as designed, since the opportunities for in-space rework, repair, and refurbishment are limited.

In consideration of these factors, the space domain has led the way in developing measures of technological stability. With the stakes so high and with so little ability to rectify problems once a spacecraft is in operational use, space developers and operators have found it necessary to ensure that only tested and stable systems and components make their way into space. Thus, NASA developed the concept of Technology Readiness Levels (TRLs) in the 1980s.[21] Based on the idea that incorporating unproven technology into critical space systems was neither safe nor cost-effective, NASA used the seven-tiered TRL process (later expanded to nine tiers) to assess objectively the maturity and stability of components prior to placing them in space systems (illustrated in Figure 1-2). This TRL concept later spread to the military and commercial worlds and developed into an entire family of assessment tools—Manufacturing Readiness Levels, Integration Readiness Levels, and Systems Readiness Levels.

As with the air domain, the R&D phase in the space domain is expensive, as is the purchase of a space vehicle itself. Unlike with aircraft, however, the O&S costs tend to be relatively low throughout the life of space systems, with operational expenses generally limited to ground station management and communication with the space vehicle. According to Defense Acquisition University, O&S costs consume 41 percent of a fixed-wing aircraft's life-cycle cost (LCC), but only 16 percent of the LCC for the average spacecraft.[22]

---

[21]Additional information on TRL definitions is available through the NASA Web site at http://esto.nasa.gov/files/TRL_definitions.pdf. Accessed June 22, 2010.

[22]Information available at https://acc.dau.mil/CommunityBrowser.aspx?id=241468. Accessed May 18, 2010.

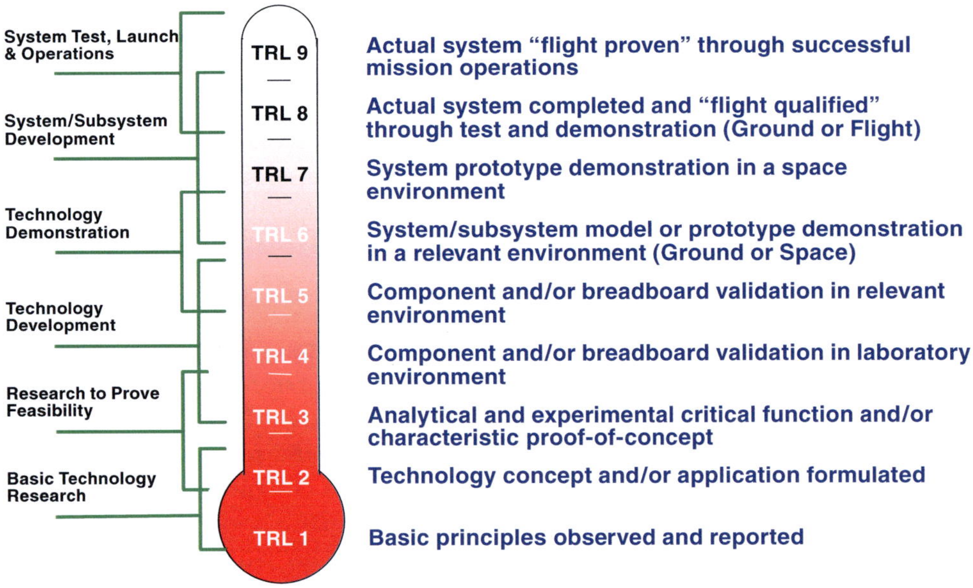

FIGURE 1-2
Technology Readiness Level (TRL) descriptions. SOURCE: NASA, modified from http://www.hq.nasa.gov/office/codeq/trl/trlchrt.pdf. Accessed June 22, 2010.

Compared to an aircraft system that can be modified to extend its life for many years, a spacecraft has a finite life on orbit, limited by the operating environment of space and the amount of fuel onboard. As a result, space systems tend to be in a constant state of acquisition. As an example, the Global Positioning System (GPS) Program, managed by the Space and Missile Systems Center, is responsible for flying the current generation of satellites on orbit, for producing the next generation of satellites, and for developing the follow-on GPS system—all at the same time.

The space domain's heavy reliance on Federally Funded Research and Development Centers (FFRDCs) is another characteristic that sets it apart from the air domain. The Aerospace Corporation has partnered with the Air Force since 1960 to provide five core technological competencies to the Air Force's space efforts:

> The Aerospace FFRDC provides scientific and engineering support for launch, space, and related ground systems. It also provides the specialized facilities and continuity of effort required for programs that often take decades to complete. This end-to-end involvement reduces development risks and costs, and allows for a high probability of mission success. The Department of Defense has identified five core competencies for the Aerospace FFRDC: launch certification, system-of-systems engineering, systems development and acquisition, process implementation, and technology application. The primary customers are the Space and Missile Systems Center of Air Force Space Command and the National Reconnaissance

Office, although work is performed for civil agencies as well as international organizations and governments in the national interest.[23]

Senior warfighters have raised concerns about the overall health of the space industrial base and its ability to meet the needs of U.S. national security. In particular, studies have pointed to inconsistent performance and reliability among third- and fourth-tier suppliers, many of which perform space contracts intermittently and cannot sustain design, engineering, and manufacturing capabilities in the absence of continual work.[24,25,26,27,28]

Cyberspace

In the early years of the Air Force's space era, there was much uncertainty about roles and missions, about organizational structure, about boundaries and policies and processes. It took decades to resolve these matters, and in fact issues still arise from time to time—an example being the recently resolved issue of whether strategic missiles logically belong to the Air Combat Command, or to the Global Strike Command, or to the Space Command.

That same level of uncertainty now characterizes the Air Force's cyberspace efforts. For example, in August 2009, a joint letter from the Secretary of the Air Force (SAF) and the Chief of Staff of the Air Force (CSAF) countermanded previous guidance and set up a new command—the 24th Air Force—as "the Air Force service component to the USCYBERCOM [United States Cyber Command], aligning authorities and responsibilities to enable seamless cyberspace operations."[29]

---

[23]Information from the Aerospace Corporation. Available at http://www.aero.org/corporation/ffrdc.html. Accessed May 18, 2010.

[24]Information from Booz Allen Hamilton. May 19, 2003. Available at www.boozallen.com/consulting/industries_article/659130. Accessed May 27, 2010.

[25]Eric R. Sterner and William B. Adkins. 2010. "R&D Can Revitalize the Space Industrial Base." *Space News*, February 22.

[26]Aerospace Industries Association. 2010. *Tipping Point: Maintaining the Health of the National Security Space Industrial Base.* September. Available at http://www.aia-aerospace.org/assets/aia_report_tipping_point.pdf. Accessed January 29, 2011.

[27]Jay DeFrank. 2006. *The National Security Space Industrial Base: Understanding and Addressing Concerns at the Sub-Prime Contractor Level.* The Space Foundation. April 4. Available at http://www.spacefoundation.org/docs/The_National_Security_Space_Industrial_Base.pdf. Accessed January 29, 2011.

[28]Defense Science Board. 2003. *Acquisition of National Security Space Programs.* Washington, D.C.: Office of the Under Secretary of Defense (Acquisition, Technology, and Logistics). Available at http://www.globalsecurity.org/space/library/report/2003/space.pdf. Accessed January 29, 2011.

[29] Available at http://www.24af.af.mil/shared/media/document/AFD-090821-046.pdf. Accessed May 18, 2010.

But as late as August 2010, much about Air Force efforts in cyberspace remained unresolved. As quoted from a 24th Air Force Web site:

> At this time, we do not yet know the full complement of wings, centers and/or other units to be assigned to 24th Air Force. The organization of the required capabilities is still being determined. . . . The exact size of 24th Air Force is unknown at this time. . . . The final numbers for Headquarters 24th Air Force are yet to be determined.[30]

Similarly, there is much yet to be learned about systems acquisition in the cyberspace domain. However, some themes can be deduced.

The first theme is the need for speed and agility in a world where threats can arise in days, or even hours. Throughout the course of this study, considerable time was devoted to learning about rapid-reaction acquisition efforts in organizations such as Lockheed Martin's Skunk Works, the Air Force's Big Safari organization, and the Joint Improvised Explosive Device Defeat Organization (JIEDDO). Common to all of these efforts was a strong sense of urgency, with program durations often measured in weeks or months rather than years and decades. But cyberspace reaction cycles are often even shorter, which for some raises the question: Is the term "major system acquisition" even relevant in the cyberspace domain?[31]

Program offices like Big Safari and JIEDDO highlight the need to keep pace with an agile and adaptive enemy; in such programs rapid acquisition processes are vital to the safeguarding of military forces and thus to the national interest. The cyberspace domain has similarly short time horizons, and these can be expected to place special demands on the acquisition of cyberspace technology (see Figure 1-3).

A second likely theme is that success in cyberspace acquisition will depend on building and rewarding a culture of innovation, and in that sense it will require more risk tolerance and failure tolerance than are commonly found in bureaucratic organizations. In 2008, the Secretary of the Air Force, the Honorable Michael Wynne, said that a new cyberspace organization would need to encourage innovation from the bottom ranks to the top:

> Calling innovation the top goal of the command, Wynne said the fledgling organization must be "more agile than any other in the Department of Defense" if it is to succeed at fighting in a climate where technology and tactics are often obsolete just months after being introduced. . . . to compete in the cyber battlefield, AFCYBER [Air Force Cyber] must be able to rapidly invent, develop and field new technologies, sometimes in a matter of weeks. To do this, the command will have to adopt a culture that encourages risk taking, especially among its young officers. . . . "Innovation will almost always come from the lower ranks," he said, "from those who have not internalized any agreed upon ways of doing things."

---

[30] Available at http://www.24af.af.mil/questions/topic.asp?id=1666. Accessed August 25, 2010.

[31] Jon Goding, Principal Engineering Fellow, Raytheon. 2010. "Improving Technology Development in Cyber: Challenges and Ideas." Presentation to the committee, June 7, 2010.

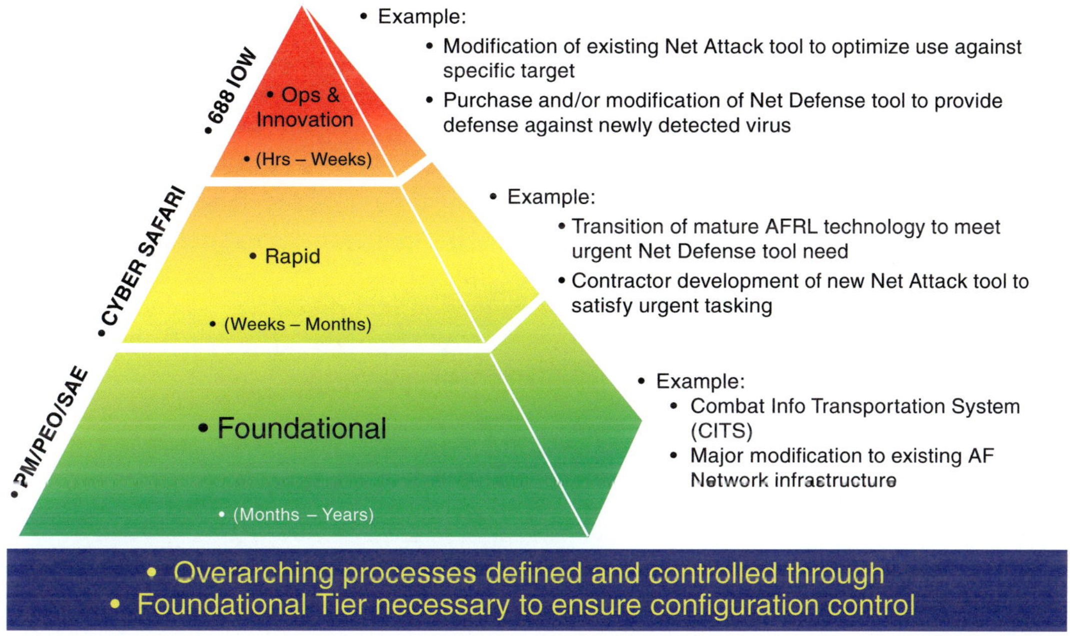

FIGURE 1-3
Cyberspace agile acquisition construct recently adopted by the United States Air Force. NOTE: Program Manager (PM), Program Executive Officer (PEO), Service Acquisition Executive (SAE), Information Operations Wing (IOW). SOURCE: Air Force Materiel Command/Electronic Systems Center.

> Wynne called on the command to foster officers who make mistakes and take risks that will lead to innovation. "Not only must you allow good ideas to percolate up, you must make your officers' careers dependent upon demonstrating innovation. Being a flawless officer in Cyber Command should lead to early retirement."[32]

A third probable characteristic of cyberspace acquisition is likely to be even closer collaboration between government, industry, and academic institutions, domestic and international. The FFRDC model discussed in the preceding subsection, on the space domain, is already a critical part of the cyberspace domain—that is, the MITRE Corporation's long-standing support of the Electronic Systems Center, and the Software Engineering Institute's support of the DoD. The need for ready access to highly specialized cyberspace expertise is very similar to the type of needs found in the space domain. Additionally, commercial industry must be brought into the collaboration, as commercial communities—for example, finan-

[32]Inside the Air Force. Available at http://integrator.hanscom.af.mil/2008/June/06262008/06262008-08.htm. Accessed May 18, 2010.

cial services, computer security—have a long history of dealing with cyberspace security, integration, and testing, and thus can contribute usefully to the Air Force's operational capability.

Technology development for air and space domains is driven by domestically based defense contractors, focusing largely on military applications. In the cyberspace world, however, the commercial market dwarfs that of the military, and both threats and resources are largely unaffected by political or geographic boundaries.[33] Cyberspace acquisitions will therefore require much tighter cooperation between technology development communities, foreign and domestic, inside and outside the military-industrial complex. A recent paper from the Air Force illustrates some of the benefits of collaborative approaches to cyberspace acquisition:

> A collaborative environment and integrated network that enables rapid reach-back into a broad and diverse array of cyber experts throughout the nation, giving the warfighter access to cutting edge technology and expertise that otherwise would be unavailable to the military. . . . A process to discover world-class cyber experts, who may be either unaware of the military cyberspace requirements or overlooked because they work for smaller, less-known firms.[34]

It may also be useful to consider that the state of the cyberspace domain has similarities to that of the nascent air domain circa 1910, or to the fledgling space world in 1960.

AIR FORCE SCIENCE AND TECHNOLOGY STRATEGIC PLANNING

During this study, technology transition was described as a "contact sport: every successful technology hand-off requires both a provider and a receiver."[35] The probability of a successful technology handoff increases when the provider and the receiver work together in a disciplined way to identify capability needs and match them to an S&T portfolio.

Approximately 30 percent of the Air Force Research Laboratory's (AFRL's) technology development efforts are "technology-push" efforts driven by technologists who perceive how an emerging technology might enable a new operational capability in advance of a stated user need—as opposed to "technology-pull" efforts, or technology development done in response to a known capability need. The

---

[33]Jon Goding, Principal Engineering Fellow, Raytheon. 2010. "Improving Technology Development in Cyber: Challenges and Ideas." Presentation to the committee, June 7, 2010.

[34]Available at http://www.docstoc.com/docs/30607347/The-Collaboration-Imperative-for-Cyberspace-Stakeholders. Accessed May 18, 2010.

[35]Michael Kuliasha, Chief Technologist, Air Force Research Laboratory. 2010. "AFRL Perspective on Improving Technology Development and Transition." Presentation to the committee, May 13, 2010.

technology-pull portion of AFRL technology development is motivated by user requirements, which originate from multiple sources, including Major Commands (MAJCOMs), Product Centers, and air logistics centers.[36]

In some cases, user needs are assessed and prioritized before they are provided to the AFRL. For example, Product Centers are again working with their warfighter partners to prioritize requirements, which are influenced, in part, by their understanding of technology enablers emerging from the Air Force, industry, and university laboratories. These needs are prioritized within a particular MAJCOM and Product Center channel, but there is no mechanism that can adequately filter and prioritize needs across the Air Force today. To its credit, the Air Force recognizes this deficiency and is taking steps to develop a more robust corporate mechanism for technology needs assessment and prioritization.[37]

THE "THREE R" FRAMEWORK

At the beginning of the study, the committee found it useful to organize its thinking around simple axiomatic principles. This resulted in a framework incorporating the following: (1) Requirements, (2) Resources, and (3) the Right People—or the "Three Rs." The framework is a concise and simple expression of unarguable criteria for successful program execution. If all three of these components are favorable, program success is possible. If any of the three is unfavorable, the program will most likely fail to deliver as expected. The framework is shown concisely in Box 1-2, and the principles are considered individually in the subsections below.

Requirements

The importance of clear, stable, feasible, and universally understood requirements has been long understood and has been validated by countless studies. Further, requirements need to be trade-off tolerant, that is, they need to be flexible enough to permit meaningful analysis of alternative solutions. Inadequately defined requirements drive program instability, through late design changes that drive cost increases and schedule slips, which in turn lead to an erosion of political support for the program. These ripples do not end at the boundary of a problematic program: As costs rise and schedules slide, the impact is transferred to other programs, and they then bear the costs imposed to save the original troubled system.

---

[36]Ibid.

[37]Ibid.

BOX 1-2
The "Three Rs"

Early in this study, the committee developed a framework, the "Three Rs," for organizing its findings and recommendations. The framework describes characteristics that, in the committee's judgment, need to be addressed fully in order for successful technology development to occur. That framework is composed of the following:

1. *Requirements*—clear, realistic, stable, trade-off tolerant, and universally understood;
2. *Resources*—adequate and stable, and including robust processes, policies, and budgets; and
3. *The Right People*—skilled, experienced, and in sufficient numbers, with stable leadership.

> Our assessment is that the current requirements process does not meet the needs of the current security environment or the standards of a successful acquisition process. Requirements take too long to develop, are derived from Joint Staff and Service views of the Combatant Commands' needs and often rest on immature technologies and overly optimistic estimates of future resource needs and availability. This fact introduces instability into the system when the lengthy and insufficiently advised requirement development process results in capabilities that do not meet warfighter needs or the capabilities that are delivered "late-to-need."[38]

A second cause of difficulty in the area of requirements is that there can be a large disconnect between what the warfighter wants—"desirements," as expressed by one presenter to the committee—and what the laws of science permit. In those cases, overly optimistic estimates early in the project life can end up requiring miracles—or worse, *sequential* miracles—in order to become reality. In the words of the *Defense Acquisition Performance Assessment Report* (called the DAPA report; commissioned by Acting Deputy Secretary of Defense Gordon England in June 2005):

> Neither the Joint Capabilities Integration and Development System nor the Services requirement development processes are well informed about the maturity of technologies that underlie achievement of the requirement or the resources necessary to realize their development. No time-phased, fiscally and technically informed capabilities development

[38]Assessment Panel of the Defense Acquisition Performance Assessment Project. 2006. *Defense Acquisition Performance Assessment Report.* A Report by the Assessment Panel of the Defense Acquisition Performance Assessment Project for the Deputy Secretary of Defense, p. 35. Available at http://www.frontline-canada.com/Defence/pdfs/DAPA-Report-web.pdf. Accessed January 29, 2011.

and divestment plan exists to guide and prioritize the development and understanding of weapon system requirements.[39]

In sum, then, a successful program requires the vigilant management of the requirements process. The Government Accountability Office (GAO) summed it up well in its 2010 report *Defense Acquisitions: Strong Leadership Is Key to Planning and Executing Stable Weapon Programs*, which studied 13 successful acquisition programs and drew lessons from those successes:

> The stable programs we studied exhibited the key elements of a sound knowledge-based business plan at program development start. These programs pursued capabilities through evolutionary or incremental acquisition strategies, had clear and well-defined requirements, leveraged mature technologies and production techniques, and established realistic cost and schedule estimates that accounted for risk. They then executed their business plans in a disciplined manner, resisting pressures for new requirements and maintaining stable funding. The programs we reviewed typically took an evolutionary acquisition approach, addressing capability needs in achievable increments that were based on well-defined requirements. To determine what was achievable, the programs invested in systems engineering resources early on and generally worked closely with industry to ensure that requirements were clearly defined. Performing this up-front requirements analysis provided the knowledge for making trade-offs and resolving performance and resource gaps by either reducing the proposed requirements or deferring them to the future. The programs were also grounded in well-understood concepts of how the weapon systems would be used.[40]

Resources

As is the case with requirements, stability of resources is essential to program success. Turbulence in any of the following areas—technology, budgets, acquisition regulation, legislation, policy, or processes—contributes to program failure, as the resulting uncertainty deprives government and industry of the ability to execute programs as planned. One key area is technological maturity. The GAO has examined the importance of technological maturity in predicting program success. In 1999, the GAO examined 23 successful technology efforts in both government and commercial projects, concluding that the use of formal approaches to assess technological stability, like the NASA-developed TRL system discussed elsewhere in this report, was crucial to program success. As stated in the 1999 GAO report:

> [D]emonstrating a high level of maturity before new technologies are incorporated into product development programs puts those programs in a better position to succeed. The

[39] Ibid., p. 36.

[40] GAO. 2010. *Defense Acquisitions: Strong Leadership Is Key to Planning and Executing Stable Weapon Programs.* Washington, D.C.: GAO, p. 16. Available at http://www.gao.gov/new.items/d10522.pdf. Accessed June 11, 2010.

> TRLs, as applied to the 23 technologies, reconciled the different maturity levels with subsequent product development experiences. They also revealed when gaps occurred between a technology's maturity and the intended product's requirements. For technologies that were successfully incorporated into a product, the gap was recognized and closed before product development began, improving the chances for successful cost and schedule outcomes. The closing of the gap was a managed result. It is a rare program that can proceed with a gap between product requirements and the maturity of key technologies and still be delivered on time and within costs.[41]

Additional emphasis on the achievement of technological maturity was mandated in the Weapon Systems Acquisition Reform Act of 2009, which requires, among many other provisions, that Major Defense Acquisition Programs (MDAPs) must, *prior to Milestone B*, carry out competitive prototyping of the system or of critical subsystems and complete their Preliminary Design Review.[42]

Similar to the need for technological maturity, there must be stability and predictability in the financial resources available to a program manager. During this study, frequent reference was made to the Valley of Death, that graveyard for technology development efforts that might survive early exploratory R&D phases, but then fall victim to a lack of funding for bridging the gap to the system development and production phases.[43] With its longer time horizons, the DoD's Planning, Programming, Budgeting, and Execution System is ill equipped to handle problems like the Valley of Death, or the sorts of rapid acquisitions that are often required today. A 2006 study from the Center for Strategic and International Studies focused on joint programs, but the message applies to all acquisition efforts:

> The current Planning, Programming, Budgeting, and Execution System (PPBES) resource allocation process is not integrated with the requirements process and does not provide sufficient resources for joint programs, especially in critical early stages of coordination between and perturbations in resource planning and requirements planning frequently result in program funding instability. Such instability increases program costs and triggers schedule slippages across DoD acquisition programs. Chronic under-funding of joint programs is endemic to the current resource allocation system.[44]

---

[41]GAO. 1999. *Best Practices: Better Management of Technology Development Can Improve Weapon System Outcomes.* GAO NSIAD-99-162. Washington, D.C.: General Accounting Office, p. 3. Available at http://www.gao.gov/archive/1999/ns99162.pdf. Accessed June 11, 2010.

[42]Weapon Systems Acquisition Reform Act of 2009 (Public Law 111-23, May 22, 2009).

[43]Dwyer Dennis, Brigadier General, Director, Intelligence and Requirements Directorate, Headquarters Air Force Materiel Command, Wright-Patterson Air Force Base, Ohio. 2010. "Development Planning." Presentation to the committee, March 31, 2010.

[44]David Scruggs, Clark Murdock, and David Berteau. 2006. *Beyond Goldwater Nichols: Department of Defense Acquisition and Planning, Programming, Budgeting, and Execution System Reform.* Washington, D.C.: Center for Strategic and International Studies. Available at http://csis.org/files/media/csis/pubs/bgnannotatedbrief.pdf. Accessed June 11, 2010.

Among the most critical resources are robust processes, from the very conception of a program. For both government and industry, well-defined and well-understood work processes in all phases of program management are essential to successful technological development. Repeatedly during this study, evidence was presented that within the Air Force some of these processes have been diluted in significant ways in the past decade and are only now beginning to be reinvigorated.

In particular, there was general agreement on the decline of the systems engineering field.[45] After a period of decline, systems engineering has been revived and received additional attention in the 2009 WSARA legislation. But once a field has been allowed to atrophy for whatever reason, the redevelopment of that capability is a long and arduous task.

A similar situation exists with the field of Development Planning. For decades, Product Centers had DP functions ("Product Center Development Planning Organizations," or, as referred to in headquarters shorthand—XRs) that worked with warfighter commands to address alternatives to meet future needs. These offices operated in the early conceptual environs, pre-program of record, to help a using command clarify its requirements, assess the feasibility of alternatives, and settle on the preferred way to meet those requirements. Often, the DP resources that worked on the early stages of a program were later used to form the initial cadre of a program office, if indeed one was ultimately established.

As with the rebirth of systems engineering, the disestablishment of Development Planning is being rectified. Product Center DP directors provided valuable input to this study, and although it is clear that their function is being reborn, it is equally obvious that a capability can be eliminated quickly by one decision but can only be revived with time and with great difficulty.

The Right People

The third critical element for a successful program is the right people, which translates to program managers and key staff with the right skills, the right experience, and in the right numbers to lead programs successfully. This category also includes the right personnel policies and "right" cultures, which can contribute to program success.

The acquisition workforce has been buffeted by change for decades. Every acquisition setback has generated a new round of "fixes," which by now have so constrained the system that it is to some a wonder that it functions at all. This workforce has been downsized, outsourced, and reorganized to the point of dis-

[45]For a thorough discussion of the importance of systems engineering, see the Kaminski report—NRC. 2008. *Pre-Milestone A and Early-Phase Systems Engineering: A Retrospective Review and Benefits for Future Air Force Systems Acquisition.* Washington, D.C.: The National Academies Press.

traction, yet there is little or no evidence to suggest that discernible improvements have resulted.

This disruption in the acquisition workforce was well recognized by those close to it. Beginning in the early 1990s, many of the best and brightest Air Force acquisition professionals chose to retire—many of them early—to take jobs with advisory and assistance services (A&AS) contractors. As these highly competent and experienced performers left, they were often not replaced, and so an enormous "bathtub" developed: Air Force acquisition specialties became understaffed, and many of the people who did remain were either in very senior oversight positions, or were very junior and, lacking mentors, very inexperienced. The middle of the force, the journeymen and junior managers, quite literally disappeared.

This was recognized by the authors of the DAPA report.[46] Released in 2006, that report accurately described the state of the acquisition workforce:

> Key Department of Defense acquisition personnel who are responsible for requirements, budget and acquisition do not have sufficient experience, tenure and training to meet current acquisition challenges. Personnel stability in these key positions is not sufficient to develop or maintain adequate understanding of programs and program issues. System engineering capability within the Department is not sufficient to develop joint architectures and interfaces, to clearly define the interdependencies of program activities, and to manage large scale integration efforts. Experience and expertise in all functional areas [have] been de-valued and contribute to a "Conspiracy of Hope" in which we understate cost, risk and technical readiness and, as a result, embark on programs that are not executable within initial estimates. This lack of experience and expertise is especially true for our program management cadre. The Department of Defense exacerbates these problems by not having an acquisition career path that provides sufficient experience and adequate incentives for advancement. The aging science and engineering workforce and declining numbers of science and engineering graduates willing to enter either industry or government will further enforce the negative impact on the Department's ability to address these concerns. With the decrease in government employees, there has been a concomitant increase in contract support with resulting loss of core competencies among government personnel.[47]

In May 2009, 3 years after the DAPA release and after being rocked by two major failed source selections in the previous year, the Air Force released its Acquisition Improvement Plan.[48] It cited five shortcomings of the acquisition process, all of

[46] Assessment Panel of the Defense Acquisition Performance Assessment Project. 2006. *Defense Acquisition Performance Assessment Report.* A Report by the Assessment Panel of the Defense Acquisition Performance Assessment Project for the Deputy Secretary of Defense. Available at http://www.frontline-canada.com/Defence/pdfs/DAPA-Report-web.pdf. Accessed January 29, 2011.

[47] Ibid., p. 29.

[48] USAF. 2009. *Acquisition Improvement Plan.* Washington, D.C.: USAF. Available at http://www.dodbuzz.com/wp-content/uploads/2009/05/acquisition-improvement-plan-4-may-09.pdf. Accessed June 11, 2010.

which pointed, in whole or in part, to failures in the human side of the acquisition enterprise:

> 1. Degraded training, experience and quantity of the acquisition workforce;
> 2. Overstated and unstable requirements that are difficult to evaluate during source selection;
> 3. Under-budgeted programs, changing of budgets without acknowledging impacts on program execution, and inadequate contractor cost discipline;
> 4. Incomplete source selection training that has lacked "lessons learned" from the current acquisition environment, and delegation of decisions on leadership and team assignments for MDAP source selections too low; and
> 5. Unclear and cumbersome internal Air Force organization for acquisition and Program Executive Officer (PEO) oversight.[49]

Clearly the two failed source selections had been a major blow to the Air Force's reputation in acquisition. The Acquisition Improvement Plan closes with a call to recapture the successes of yesterday:

> We will develop, shape, and size our workforce, and ensure adequate and continuous training of our acquisition, financial management, and requirements generation professionals. In so doing we will re-establish the acquisition excellence in the Department of the Air Force that effectively delivered the Intercontinental Ballistic Missile; the early reconnaissance, weather, and communication satellites; the long-range bombers like the venerable B-52; and fighters like the ground-breaking F-117A. . . .[50]

Those steps are under way. Evidence was presented during this study indicating that expedited hiring was being used to fill empty positions, in accordance with the DAPA recommendations.[51] However, the redevelopment of a skilled and experienced workforce is in some ways reminiscent of the challenges facing those seeking to reinvigorate systems engineering or Development Planning: Similar to what was seen with those critical processes, a skilled workforce can shrink quickly, yet will take decades or more to rebuild and mature.

REPORT ORGANIZATION

The remainder of this report is structured as follows, to correspond to the four main paragraphs of the statement of task. Chapter 2, "The Current State of

---

[49]Ibid.

[50]Ibid., p. 14.

[51]Assessment Panel of the Defense Acquisition Performance Assessment Project. 2006. *Defense Acquisition Performance Assessment Report.* A Report by the Assessment Panel of the Defense Acquisition Performance Assessment Project for the Deputy Secretary of Defense. Available at https://acc.dau.mil/CommunityBrowser.aspx?id=18554. Accessed June 11, 2010.

the Air Force's Acquisition Policies, Processes, and Workforce," addresses the first paragraph of the statement of task. Chapter 3, "Government and Industry Best Practices," addresses the third paragraph of the statement of task. Chapter 4, "The Recommended Path Forward," responds to the second and fourth paragraphs of the statement of task. Importantly, the committee chose to present its findings in Chapters 2 and 3, and the associated recommendations (plus the reiterated relevant findings from the earlier chapters) are consolidated in Chapter 4. Finally, Appendixes C and D provide background information related to the subjects addressed in Chapter 2.

2

The Current State of the Air Force's Acquisition Policies, Processes, and Workforce

From its founding, the United States Air Force (USAF) was based on the premise that gaining and maintaining technological supremacy were essential to its combat success. This technological superiority provided great benefit for both the Air Force and the nation. Comparing its current efforts to its past achievements and successes, the Air Force today finds itself struggling to successfully field new technology in its weapons systems on schedule and within budget. According to a recent independent assessment:

> [T]he AF has experienced a number of symptoms that indicate problems with its acquisition system and processes [that bring new technology into operational use]. Some of the most pressing of these symptoms have been: (a) numerous cost-schedule-performance issues; (b) numerous Nunn-McCurdy unit cost breaches; (c) increased time to bring major systems to the field; and (d) successful protests by contractors on major programs.[1]

This chapter identifies key issues that affect the ability of the Air Force to specify, develop, test, and insert new technology into its major new systems.

CURRENT AND HISTORICAL POLICIES AND PROCESSES RELATED TO TECHNOLOGY DEVELOPMENT

The statement of task for this study required the committee to "examine appropriate current or historical Department of Defense (DoD) policies and processes,

[1]Gary E. Christle, Danny M. Davis, and Gene H. Porter. 2009. *CNA Independent Assessment. Air Force Acquisition: Return to Excellence.* Alexandria, Va.: CNA Analysis & Solutions.

including the Planning, Programming, Budgeting, and Execution System, DoD Instruction 5000.02, the Air Force Acquisition Improvement Plan, the Joint Capabilities Integration and Development System, and DoD and Air Force competitive prototyping policies to comprehend their impact on the execution of pre-program of record technology development efforts" (for the full statement of task, see Box 1-1 in Chapter 1). The descriptions in the following subsections summarize the above policies and processes. Table 2-1 lists the current policies and processes, as well as their unintended consequences or shortfalls. A more comprehensive discussion of these policies and processes is provided in Appendix C.

Planning, Programming, Budgeting, and Execution System

In the planning phase of the Planning, Programming, Budgeting, and Execution System (PPBES), the Office of the Secretary of Defense (OSD) and the Joint Staff collaboratively articulate national defense policies and military strategy in the Strategic Planning Guidance (SPG).[2] The result is a set of budget-conscious priorities for program development (military force modernization, readiness, and sustainability; and supporting business processes and infrastructure), which is promulgated in the Joint Programming Guidance (JPG).

The next phase of the PPBES, programming, begins with the writing of the Air Force Program Objective Memorandum (POM). The POM balances program budgets as set down in the JPG. The third phase, budgeting, happens concurrently with the programming phase. Each DoD department and agency submits its budget estimate with its POM. The DoD departments and agencies then convert their program budgets into the congressional appropriation structure format and submit them, along with justification. The budget forecasts only the next 2 years, but with more detail than the POM. Execution is the responsibility of the individual services. The Planning, Programming, Budgeting, and Execution (PPBE) process is a very high level strategic process, and as such it addresses preacquisition technology development only indirectly.[3]

Department of Defense Instruction 5000.02

Although DoD Instruction 5000.02 discusses the preacquisition phase, it provides little "how-to" guidance, nor does it provide any formal direction regarding

---

[2]Abstracted from the Defense Acquisition Web site. Available at https://dap.dau.mil/aphome/ppbe/Pages/Default.aspx. Accessed August 10, 2010.

[3]Thomas Thurston, Program Manager, PPBE Processes and Training Programs, Science Applications International Corporation (SAIC). 2010. "PPBE Executive Training." Presentation to the committee, April 21, 2010.

TABLE 2-1 Current DoD Policies and Processes and Their Unintended Consequences

Policy/Process, Date Instituted	Office	Formal Policy?	Formal Process?	Technology Development Addressed?	Results Metric?	Intent	Potential Unintended Consequences or Shortfalls
PPBES, May 22, 1984	OUSD (C)	Yes	Yes	No	No	Institute a rigorous budgetary process	Technology development not adequately addressed
JCIDS, June 2003	CJCS/J8	Yes	Yes	No	No	Include COCOM priorities in modernization efforts	Technology development "how-to's" not addressed
Section 852 of the 2008 NDAA, January 28, 2008	Congress	Yes	Yes	No	Yes	Recruit, retain, educate, and train DoD acquisition workforce	Recruiting focused on systems engineers, contracting officers, and cost estimators; insourcing of work done by support contractors; lack of flexibility; increased program office cost; recruiting in high-cost areas; loss of quick-response specialized domain expertise
CP, December 8, 2008	OUSD (AT&L)	Yes	Yes	Yes	Yes	Preserve competition and gain the benefits of maximum innovation	Preserves industry design teams; all critical core subsystems not assessed; significant changes to requirements after prototype development
DoDI 5000.02, December 8, 2008	OUSD (AT&L)	Yes	Yes	Yes	Yes	Define acquisition process and policy	Technology development oversight may become onerous
AIP, May 2009	SAF/AQ	Yes	Yes	No	No	Revitalize Air Force acquisition process	Fails to address technology development
WSARA, May 22, 2009	Congress	Yes	Yes	Yes	No	Reform and improve DoD acquisition and technology development	Additional nonproductive oversight

NOTE: Acronyms are defined in the list in the front matter of this report.

the employment of DoD Instruction 5000.02, the training of the acquisition workforce, or the assessment of acquisition workforce skills.[4] One significant change incorporated in DoD Instruction 5000.02 is the increased emphasis on technology development and maturation.[5] Previously, in DoD Instruction 5000.2, technology development was part of the pre-systems acquisition phase, focused more on concept exploration and Analysis of Alternatives (AoA).[6] Many of the technology transition objectives and mechanisms cited in DoD Instruction 5000.2 have been retained in the current DoD Instruction 5000.02, but the pre-systems acquisition phase between Milestone A and Milestone B is now focused on reducing technology risk prior to contracting for Engineering and Manufacturing Development (EMD).

The entrance criteria for technology development at Milestone A now include the requirement for a technology development strategy (TDS) and full funding for the technology development phase of the acquisition program. The new DoD Instruction 5000.02 includes two additional mandates that need to be addressed in future Air Force acquisition programs. The instruction requires the acquisition authority to fund two or more competing prototypes of the system or key system elements and, when consistent with technology development phase objectives, to accomplish a Preliminary Design Review prior to Milestone B.[7]

Prototyping can certainly reduce risk if the prototype is truly representative of the production concept in function, performance characteristics, and emergent properties. Likewise, early design review will bolster confidence if it is accomplished at an appropriate level of detail and analytical rigor. The challenge to the acquisition community going forward is to take on these two mandates in a meaningful way. It may be that in this context "one size can't fit all." The funding and schedule allocations required to accomplish meaningful prototypes of more complex systems (e.g., aircraft and spacecraft) or systems of systems (e.g., battlespace management information technology, multiple autonomous systems) are likely to reach a point of diminishing returns.

[4]DoD. 2008. *Department of Defense Instruction. Subject: Operation of the Defense Acquisition System.* 5000.02. Washington, D.C.: DoD. Available at http://www.dtic.mil/whs/directives/corres/pdf/500002p.pdf. Accessed August 11, 2010.

[5]Ibid.

[6]USAF. 2008. *Analysis of Alternatives (AoA) Handbook: A Practical Guide to Analysis of Alternatives.* Kirtland Air Force Base, N.Mex.: Air Force Materiel Command (AFMC) Office of Aerospace Studies. Available at http://www.oas.kirtland.af.mil/AoAHandbook/AoA%20Handbook%20Final.pdf. Accessed June 10, 2010.

[7]DoD. 2008. *Department of Defense Instruction. Subject: Operation of the Defense Acquisition System.* 5000.02. Washington, D.C.: DoD. Available at http://www.dtic.mil/whs/directives/corres/pdf/500002p.pdf. Accessed August 11, 2010.

Air Force Acquisition Improvement Plan

The only mention of technology in the Air Force Acquisition Improvement Plan (AIP) is a caution to warfighters to "resist the temptation to pursue high risk requirements that are too costly and take too long to deliver in favor of an incremental acquisition strategy that delivers most, if not all, requirements in the initial model with improvements added as technology matures. . . ."[8] It makes no mention of mechanisms by which technology will be developed and matured in the preacquisition phase, or, for that matter, in any phase of the acquisition life cycle.

Joint Capabilities Integration and Development System

Chairman of the Joint Chiefs of Staff Instruction (CJCSI) 3170.01G describes the need for requirement support in concert with the resourcing and acquisition processes, to support the preacquisition program phase as well as Milestone B and beyond.[9] However, CJCSI 3170.01G describes the need and directs the strong involvement of the requirements community in the preacquisition phase, but the Joint Capabilities Integration and Development System (JCIDS) manual written to implement CJCSI 3701.01G provides insufficient "how-to" guidance on integration of the requirement into the acquisition and resource processes.

Competitive Prototyping

Although DoD and Air Force competitive prototyping policies and processes do focus on the preacquisition program phase, the DoD documentation does not provide clear methodologies, is silent on workforce training policies, and offers few metrics for tracking progress. The Air Force competitive prototyping policy, Air Force Instruction (AFI) 63-101, however, does provide processes, methodologies, and some measures for tracking progress.[10] One shortcoming of AFI 63-101 is that it lacks a waiver process, whereas the Weapon Systems Acquisition Reform Act of 2009 (WSARA; Public Law 111-23) and DoD policy allow for waivers.

---

[8]USAF. 2009. *Acquisition Improvement Plan.* Washington, D.C.: Headquarters United States Air Force. May 4. Available at http://images.dodbuzz.com/wp-content/uploads/2009/05/acquisition-improvement-plan-4-may-09.pdf. Accessed January 29, 2011. p. 6.

[9]Joint Chiefs of Staff. 2009. *Joint Capabilities Integration and Development System. Chairman of the Joint Chiefs of Staff Instruction.* CJCSI 3170.01G. March 1. Washington, D.C.: Joint Chiefs of Staff. Available at http://www.dtic.mil/cjcs_directives/cdata/unlimit/3170_01.pdf. Accessed August 10, 2010.

[10]USAF. 2010. *Air Force Guidance Memorandum to AFI 63-101: Acquisition and Sustainment Life Cycle Management.* Washington, D.C.: Department of Defense. Available at http://www.af.mil/shared/media/epubs/AFI63-101.pdf. Accessed August 11, 2010.

FINDING 2-1
The Air Force competitive prototyping policy, AFI 63-101, lacks a waiver process for competitive prototyping.

HISTORICAL GOVERNANCE RELATED TO TECHNOLOGY DEVELOPMENT

Recently, several high-profile studies[11,12,13,14] have discussed, at least tangentially, technology development. In addition, various policies, processes, and laws have been enacted over the years addressing technology development.[15,16,17,18,19] Table 2-2 provides a summary of the committee's assessment of the policies and processes discussed above and highlights specific unintended consequences or shortfalls in the context of technology development.

THE TRUST "DEATH SPIRAL"

As discussed in Chapter 1, in the subsection "The Right People," forces external and internal to the technology development and acquisition processes have

[11]NRC. 2008. *Pre-Milestone A and Early-Phase Systems Engineering: A Retrospective Review and Benefits for Future Air Force Systems Acquisition.* Washington, D.C.: The National Academies Press.

[12]Assessment Panel of the Defense Acquisition Performance Assessment Project. 2006. *Defense Acquisition Performance Assessment Report.* A Report by the Assessment Panel of the Defense Acquisition Performance Assessment Project for the Deputy Secretary of Defense. Available at https://acc.dau.mil/CommunityBrowser.aspx?id=18554. Accessed June 10, 2010.

[13]Gary E. Christle, Danny M. Davis, and Gene H. Porter. 2009. *CNA Independent Assessment. Air Force Acquisition: Return to Excellence.* Alexandria, Va.: CNA Analysis & Solutions.

[14]Business Executives for National Security. 2009. *Getting to Best: Reforming the Defense Acquisition Enterprise. A Business Imperative for Change from the Task Force on Defense Acquisition Law and Oversight.* Available at http://www.bens.org/mis_support/Reforming%20the%20Defense.pdf. Accessed June 10, 2010.

[15]USAF. 2008. *Analysis of Alternatives (AoA) Handbook: A Practical Guide to Analysis of Alternatives.* Kirtland Air Force Base, N.Mex.: Air Force Materiel Command (AFMC) Office of Aerospace Studies. Available at http://www.oas.kirtland.af.mil/AoAHandbook/AoA%20Handbook%20Final.pdf. Accessed June 10, 2010.

[16]Office of History Headquarters, Air Force Systems Command. 1979. *History of the Air Force Systems Command—Calendar Year 1978.* October 15. Washington, D.C.: U.S. Air Force.

[17]The Defense Acquisition Workforce Improvement Act (Public Law 101-510). More information is available at http://www.dau.mil/pubscats/PubsCats/acker/garci.pdf. Accessed August 13, 2010.

[18]The Department of Defense Acquisition Workforce Development Fund under Title X. More information is available at http://www.law.cornell.edu/uscode/10/usc_sec_10_00001705----000-.html. Accessed August 13, 2010.

[19]The Weapon Systems Acquisition Reform Act of 2009 (Public Law 111-23). More information is available at http://www.ndia.org/Advocacy/PolicyPublicationsResources/Documents/WSARA-Public-Law-111-23.pdf. Accessed August 13, 2010.

TABLE 2-2 Historical DoD Policies and Processes and Their Unintended Consequences

Policy/Process, Date Instituted	Office	Formal Policy?	Formal Process?	Technology Development Addressed?	Results Metric?	Intent	Unintended Consequences
PPBS, Early 1960s	OUSD (C)	Yes	Yes	No	No	Institute a rigorous budgetary process	Failure to achieve much of the intended benefit
Vanguard, Late 1970s	AFSC/XR	Yes	Yes	Yes	Yes	Establish a rigorous, comprehensive Development Planning process. Provide the Air Force an annual, cooperatively approved strategic and tactical roadmap of Air Force capabilities and provided rigorous justification for it in budget requests	Congressional staff saw the results as a threat to budget discipline.
Goldwater-Nichols Department of Defense Reorganization Act, October 1, 1986	Congress	Yes	Yes	No	No	Fix inter-service rivalries; provide for shared procurement; shift acquisition responsibility to the Secretariat from the military chain of command.	Dual chain of command for programs, separated requirements and money from acquisition chain of command, adversely impacted programs, reduced trust, increased rules and regulations
Defense Acquisition Workforce Improvement Act, November 5, 1990	Congress	Yes	Yes	Yes	Yes	Improve acquisition workforce	While attempting to enhance the professionalism of the acquisition corps, it has made these professionals more insular from the warfighter and other service professionals.
DoD Instruction 5000.2, April 5, 2002	OUSD (AT&L)	Yes	Yes	Yes	No	Instill acquisition discipline; provide independent oversight	Growth in the complexity of major program management
Program Budget Decision 720, 2005	OSD	No	Yes	No	No	Reduce acquisition and technology workforce	Smaller acquisition workforce, lost skills and expertise, loss of trust, more oversight

NOTE: Acronyms are defined in the list in the front matter of this report.

caused a major reduction in the numbers of people at the execution level, and large numbers of experienced, motivated, and skilled acquisition and technology professionals have left the government workforce, either voluntarily or involuntarily. In the words of the Commander of Air Force Materiel Command: "We have lost the ability to grade our contractors' homework."[20]

Additionally, the Air Force's increased use of Total System Performance Responsibility (TSPR) as a contracting strategy resulted in a substantial loss of in-house technical expertise, as well as additional reductions of nontechnical personnel supporting the acquisition process. For example, Figures 2-1 and 2-2 illustrate the significant decline in the engineering workforce at both the Air Force Materiel Command (AFMC) and the Air Force Space and Missile Systems Center (SMC).[21,22] An additional complicating factor was a congressional cap on supporting Federally Funded Research and Development Centers (FFRDCs) that substantially reduced the outside technical support provided to some, but not all, of the Product Centers.[23] During a period of increased programmatic and technical complexity, there has been a significant loss of the most experienced members of the acquisition workforce, without an adequate replacement pipeline. Additionally, without the necessary management emphasis, there has been "a systematic failure to update specifications, standards, and handbooks" that are essential to a successful acquisition system.[24,25,26]

Such significant personnel losses, combined with the atrophy of relevant guidance and documentation, have contributed to technology development and acqui-

[20]Donald Hoffman, General, Commander, Air Force Materiel Command, USAF. Personal communication to the committee, July 15, 2010.

[21]Vincent Russo, Executive Director, Aeronautical Systems Center, United States Air Force. 2003. "An Overlooked Asset: The Defense Civilian Workforce." Statement before the Committee on Governmental Affairs Subcommittee on Oversight of Government Management, United States Senate.

[22]Dwyer Dennis, Brigadier General, Director, Intelligence and Requirements Directorate, Headquarters Air Force Materiel Command, Wright-Patterson Air Force.

[23]GAO. 1996. *Federally Funded R&D Centers: Issues Relating to the Management of DoD-Sponsored Centers.* Washington, D.C.: GAO. Available at http://www.gao.gov/archive/1996/ns96112.pdf. Accessed July 22, 2010.

[24]Arthur Huber, Colonel, Vice Commander, Aeronautical Systems Center, Wright-Patterson Air Force Base; and Gerald Freisthler, Executive Director, Aeronautical Systems Center, Wright-Patterson Air Force Base. 2010. "Aeronautical Systems Center Involvement in Applied Technology Councils." Presentation to the committee, June 1, 2010.

[25]GAO. 2005. *Information Technology: DoD's Acquisition Policies and Guidance Need to Incorporate Additional Best Practices and Controls.* Washington, D.C.: GAO. Available at http://www.gao.gov/new.items/d04722.pdf. Accessed July 22, 2010.

[26]Information on workforce decline is also found in GAO. 2009. *Defense Critical Infrastructure: Actions Needed to Improve the Consistency, Reliability, and Usefulness of DoD's Tier 1 Task Critical Asset List.* Washington, D.C.: GAO. Available at http://www.roa.org/site/DocServer/GAO_Defense_Infrastructure_17_Jul_09.pdf?docID=19801. Accessed July 22, 2010.

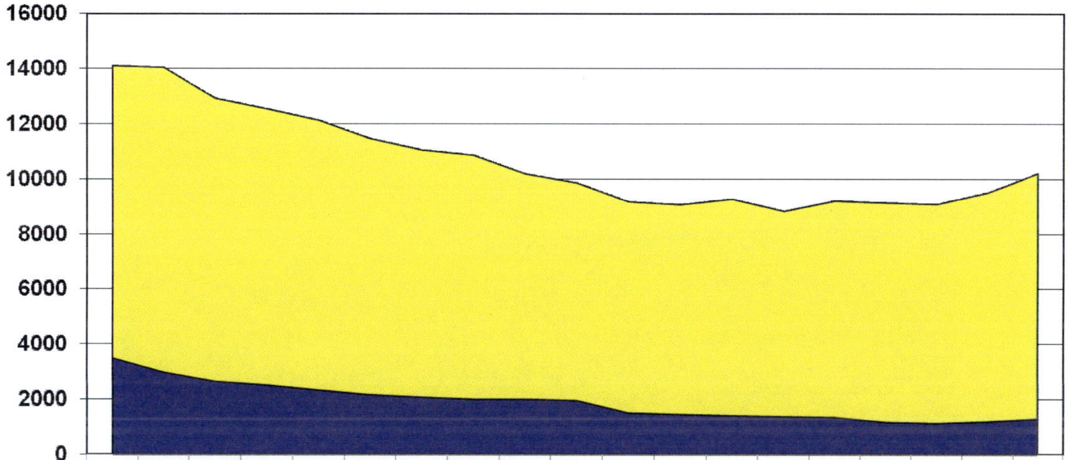

	FY92	FY93	FY94	FY95	FY96	FY97	FY98	FY99	FY00	FY01	FY02	FY03	FY04	FY05	FY06	FY07	FY08	FY09	FY10
CIV	10613	11072	10285	10028	9799	9307	8978	8865	8183	7907	7671	7618	7862	7453	7860	7967	7950	8312	8914
MIL	3481	2965	2636	2509	2323	2154	2066	1995	1996	1946	1498	1444	1402	1369	1336	1164	1126	1189	1289

Authorizations with a Scientists (61S) or Developmental Engineering (62E) AFSC

FIGURE 2-1
Summary of Air Force Materiel Command (AFMC) science and engineering workforce authorizations. SOURCE: Steven Butler, Former Executive Director, Air Force Materiel Command, USAF. Personal communication with the committee, August 24, 2010.

sition failures. Those failures have caused the Congress, the Office of the Secretary of Defense, and higher headquarters in the Air Force to lose confidence in the executing organizations, and as a remedy, additional layers of oversight have been added.[27,28,29] Although oversight can be value-added when conducted by knowledgeable people in a constructive manner, the very nature of such oversight tends to be based on distrust rather than on trust.

Ever-increasing oversight resulting from this lack of trust has greatly added to the workload of the people at the execution level, further reducing the time

[27]Dwyer Dennis, Brigadier General, Director, Intelligence and Requirements Directorate, Headquarters Air Force Materiel Command, Wright-Patterson Air Force Base, Ohio. 2010. "Development Planning." Presentation to the committee, March 31, 2010.

[28]Steven Walker, Senior Executive Service, Deputy Assistant Secretary of the Air Force for Science, Technology and Engineering, Washington, D.C., SAF/AQR. 2010. "Evaluation of Air Force Preacquisition Technology Development." Presentation to the committee, March 30, 2010.

[29]Michael Sullivan, Director, Acquisition and Sourcing Management Team, GAO. 2010. "Survey of GAO Studies and Findings." Presentation to the committee, April 21, 2010.

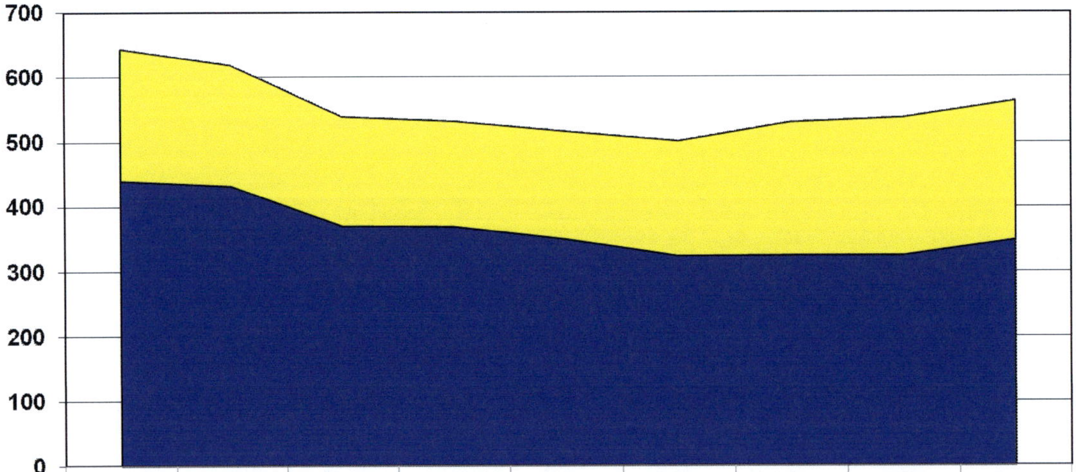

	FY02	FY03	FY04	FY05	FY06	FY07	FY08	FY09	FY10
CIV	204	187	169	163	167	178	206	213	215
MIL	440	432	370	369	349	323	324	324	348

Authorizations with a Scientists (61S) or Developmental Engineering (62E) AFSC

FIGURE 2-2
Summary of Air Force Space Command (AFSPC) science and engineering workforce authorizations. NOTE: The Space and Missile Systems Center (SMC) transitioned from the Air Force Materiel Command to AFSPC (starting in fiscal year 2002). SOURCE: Donald Wussler, Colonel, Director, Development Planning, Space and Missile Systems Center, USAF. Personal communication with the committee, August 27, 2010.

available to them to manage technology development and acquisition programs responsibly.[30] One result of this declining trust has been the passage of WSARA, directing independent assessments of Technology Readiness Levels (TRLs) at the OSD-level.

A remedy—the reconstitution of an experienced and capable Air Force acquisition workforce that would include program managers, financial and contracting personnel, testers, and evaluators, as well as the technical staff to support program offices—has been initiated, but it will take much time and effort. There has been considerable emphasis placed recently on the reinvigoration of systems engineering; however, organic subject-matter experts within each of the domains are of equal importance. After two decades of atrophy, pipelines for the accession and

[30]Dwyer Dennis, Brigadier General, Director, Intelligence and Requirements Directorate, Headquarters Air Force Materiel Command, Wright-Patterson Air Force Base, Ohio. 2010. "Development Planning." Presentation to the committee, March 31, 2010.

development of technically skilled and broadly experienced military and Civil Service personnel must be reestablished. This will require exceptional constancy and consistency of purpose from Air Force leadership.

Similarly, the reestablishment of trust will take time and is dependent on the redevelopment of a capable and experienced workforce, with the wisdom and discipline necessary to avoid the numerous acquisition problems that have plagued the process over the past 20 years. If appropriate and effective corrective action to rebuild the workforce is not taken, the result will be worsening levels of performance and an ever-more-hostile environment in which technology development and acquisition are conducted. Such a cycle can result in an ever-worsening "Death Spiral," in which lack of trust and the resultant excessive independent oversight exacerbate programmatic instability, as shown in Figure 2-3.

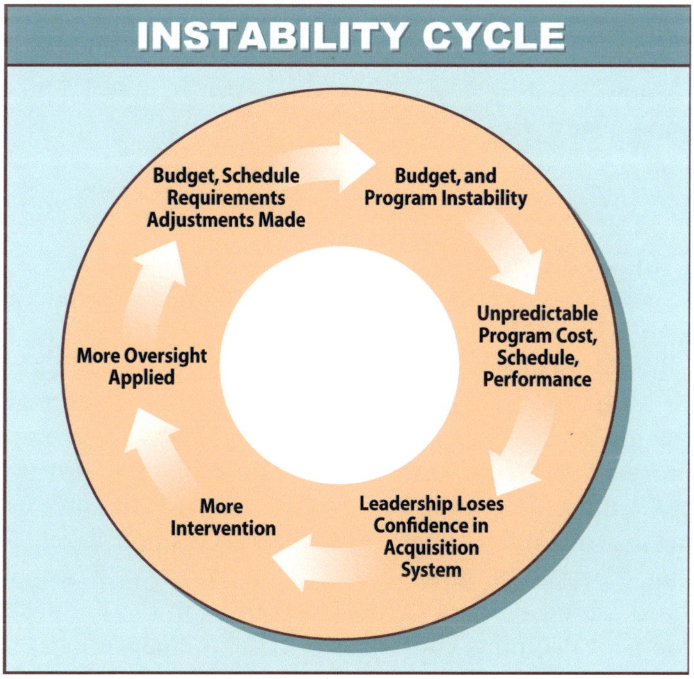

FIGURE 2-3
The management and oversight systems of the Department of Defense generate significant program instability. SOURCE: Reprinted from Assessment Panel of the Defense Acquisition Performance Assessment Project. 2006. *Defense Acquisition Performance Assessment Report.* A Report by the Assessment Panel of the Defense Acquisition Performance Assessment Project for the Deputy Secretary of Defense. Available at https://acc.dau.mil/CommunityBrowser.aspx?id=18554. Accessed June 10, 2010.

FINDING 2-2
Lack of trust and increasing oversight of Air Force technology development and acquisition by the Congress, OSD, and Air Staff are making successful program execution ever more difficult.

THE "THREE R" FRAMEWORK

Building on Chapter 1, this chapter uses the "Three R" framework—that is, (1) Requirements, (2) Resources, and (3) the Right People—to track the Air Force's preacquisition state in general and its experience with technology development specifically. Under "Requirements," there are three focus areas: namely, the loss of Development Planning (DP), the decline in the effectiveness of Applied Technology Councils (ATCs), and the need to ensure that technology is mature enough to be incorporated in acquisition programs. Under "Resources," the discussion emphasizes the lack of an Air Force-level science and technology (S&T) strategy. And under "The Right People," the point is made that the workforce within the Air Force responsible for acquisition and technology development has atrophied to the point that it is now insufficient in both quality—that is, meaningful and relevant experience—and quantity.

Requirements

Previous studies suggest that the Air Force needs to do more effective planning in the earliest stages of programs, when ultimate cost, schedule, and technical performance are most malleable and thus most readily influenced. Recently, the report from the National Research Council referred to as the Kaminski report addressed this aspect directly, highlighting the need for systems engineering and the importance of the role that systems engineering plays in the major systems acquisition process.[31] It also persuasively made the case for a return to the days of Development Planning, describing how prior to 1990 the Air Force used Development Planning to assess and integrate the various acquisition stakeholder communities, including especially combat commands, the Air Force Research Laboratory (AFRL), and acquisition Product Centers. According to the Kaminski report, the use of Development Planning, coupled with systems engineering, resulted in the delivery of needed capability to the warfighter in a timely and affordable manner.[32]

In addition to Development Planning, there exist two other significant tools in the quest for clear, realistic, trade-off-tolerant, stable, and universally understood requirements. The first of these other tools consists of the once-effective ATCs, in

[31] NRC. 2008. *Pre-Milestone A and Early-Phase Systems Engineering: A Retrospective Review and Benefits for Future Air Force Systems Acquisition.* Washington, D.C.: The National Academies Press.

[32] Ibid.

which warfighting commands, acquisition and logistics organizations, and laboratories managed the linkages between operational requirements, technology development, and systems acquisition—with the added benefit of the interpersonal relationships that developed, as well as the face-to-face communications which ensued. The second tool is the establishment and disciplined use of measures of technological readiness—that is, Technology Readiness Assessments (TRAs)—so that only when a technology is well defined and demonstrated does it make the transition from the laboratory world to become part of a major system acquisition program. Each of these—Development Planning, Applied Technology Councils, and Technology Readiness Assessments—is discussed below.

The Fall—and Rise—of Development Planning

The need for clear, realistic, and stable requirements has long been understood. Shifting requirements are key drivers of program instability, causing late design changes that drive cost increases and schedule slippage, which in turn can lead to an erosion of political support for a program. To avoid these sources of program turbulence, there needs to be a clear understanding of requirements on the part of all stakeholders. This was in fact the role of Development Planning, in which an experienced cadre of Product Center acquisitions experts, knowledgeable of both warfighter requirements and the state of relevant technologies, facilitated this clear and mutual understanding of what was needed, and—equally important—what was possible (as illustrated in Figure 2-4).

Development Planning served the Air Force well. Until its demise a decade ago, many major weapons system acquisition programs were conceptualized, their requirements developed and refined, their technologies selected and matured, and their life-cycle costs accurately projected. Funding for Development Planning under Program Element 65808 fluctuated with the times, until it was eventually eliminated by Congress in 2001, with much of the DP expertise being scattered and eventually lost.[33] The resulting absence of experienced staff members and mature processes did great damage to the understanding and development of requirements, while the past successes of Development Planning soon became mere memories in the minds of retirees and historians. It took only a brief time for the elimination of Development Planning to show up in major acquisition program failures, and numerous presenters to the committee pointed to the abandonment of Development Planning as a major contributing factor in the decline of Air Force acquisition excellence. One example of this decline is the E-10A Multisensor Command and Control aircraft that became a full-fledged program of record without any real AoA

[33]For more information on Program Element 65808, see NRC. 2008. *Pre-Milestone A and Early-Phase Systems Engineering: A Retrospective Review and Benefits for Future Air Force Systems Acquisition.* Washington, D.C.: The National Academies Press.

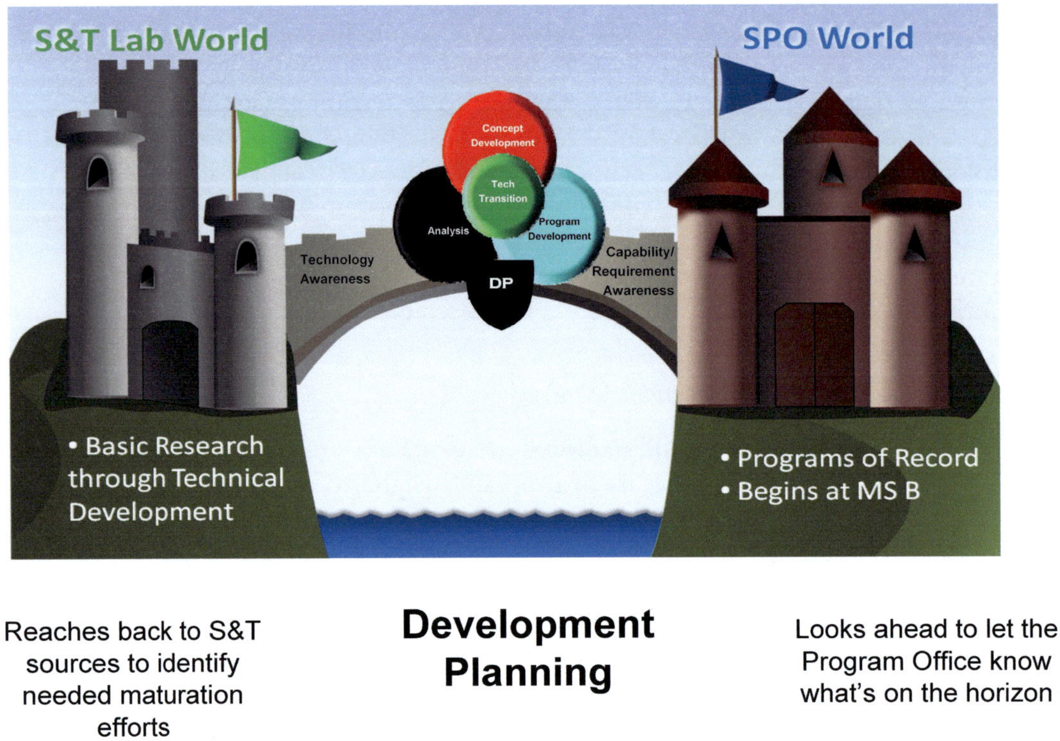

FIGURE 2-4
The role of the Development Planning organization is to reach back into the science and technology (S&T) world to identify and assess the maturity level of technology necessary to meet operational requirements and to inform a System Program Office (SPO) about new technologies on the horizon. SOURCE: Dwyer Dennis, Brigadier General, Director, Intelligence and Requirements Directorate, Headquarters Air Force Materiel Command, Wright-Patterson Air Force Base, Ohio. 2010. "Development Planning." Presentation to the committee, March 31, 2010.

and that was also eventually canceled. The Kaminski report also addressed Development Planning and its important relationship to systems engineering, stating that the role of Development Planning was as follows:

> [T]o employ various tools and techniques to define defense strategies, identify gaps in accomplishing those strategies, define concepts to address the gaps, use modeling and simulations or prototyping as ways to refine and test concepts, and provide early systems requirements to the systems developers for specific programs. Inherent in this role was the ability to understand the state of the art of the technical possibilities available from technology centers (laboratories, universities, industry, and so on), as well as to understand the needs of the user community (warfighters). These are all key attributes of a good pre-Milestone A systems engineering process. Successful programs discussed in Chapter 2 as "best practices" (e.g., C-5 and B-2) were originated during the "development planning" era.[34]

[34]Ibid., p. 21.

It is clear that in order to provide capability to the warfighter more effectively and affordably, the Air Force needs to revive Development Planning, and actions are being taken to do that (as shown in Figure 2-5). Still, there are reasons for concern. Although current efforts may be a good start, it became apparent to the committee that the Air Force would benefit from having a better understanding of previous DP processes that had been successful for so long. Some presenters to the committee appeared unaware of the Kaminski report's recommendations regarding Development Planning. In addition, all presentations from Product Center DP chiefs indicated that their organizations suffered from high workload, limited personnel, and inadequate funding.

The Air Force is developing and reinvigorating its DP process. However, as in all organizations, Air Force decision makers would profit from a clearer understanding of their own past. There is very little new in the management of technology development: Indeed, the important lessons of technology development and acquisition management have been learned previously by the Air Force and, regrettably, many seem to have been forgotten. One such example involves an earlier Air Force DP process called Vanguard. The struggles—and the successes—of Vanguard 30 years ago closely parallel the challenges facing the Air Force today. Appendix D provides background information on the Vanguard process.

The Decline of Applied Technology Councils

At one time, Applied Technology Councils were an effective tool for integrating warfighter requirements with acquisition priorities and technology maturation efforts. Hosted by Product Center commanders, ATCs were held quarterly and were attended by senior-level warfighters, top laboratory management, and high-level acquisition leaders. Combat commanders made clear their operational requirements, laboratory leaders explained what was feasible technologically, and the acquisition community set forth programmatic plans for matching requirements with new systems or subsystems. Priorities were established, funding was committed, and plans were made to transition technologies from the S&T world, across the Valley of Death, to operational success (as shown in Figure 2-6).

In some cases, ATCs remain important tools for managing the requirements-technology-acquisition interface. Through the research of the committee it became clear, however, that ATCs have, in some arenas, been allowed to deteriorate past the point of usefulness. The causes were many: New commanders sometimes had other priorities, while other, overtasked acquisition leaders began to let the intervals between ATCs grow, first to semiannual intervals, then to annual intervals, and then beyond that. When asked how often ATCs were held, one respondent told the committee, "Annually. . . . But sometimes we cancel them." The staffs of some participating organizations soon required multiple pre-briefings, adding bureaucracy to the process and arguably watering down the frank dialogue. Eventually,

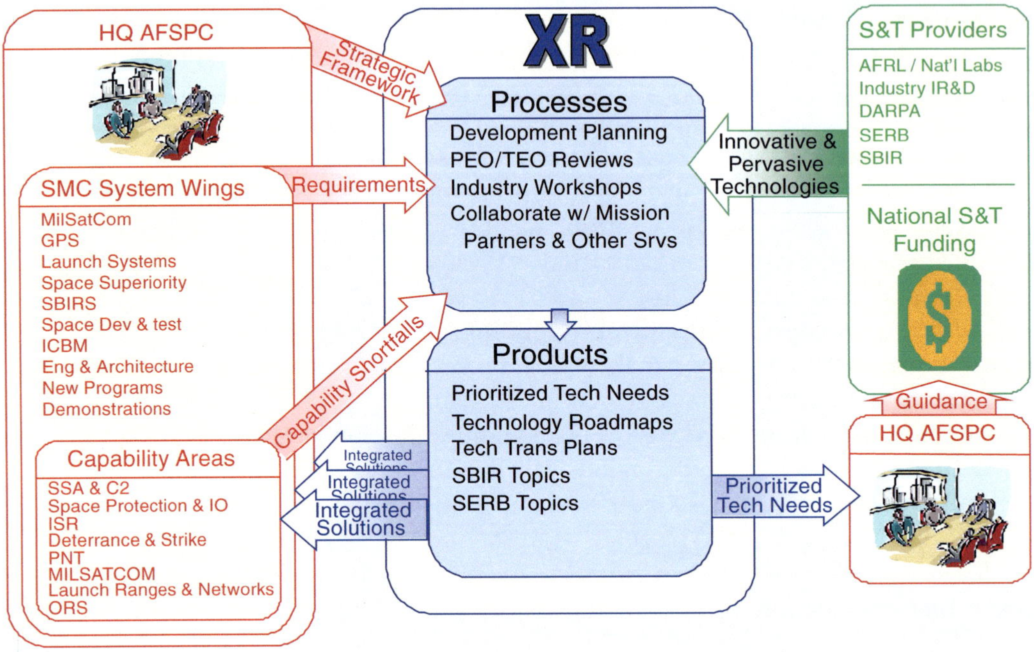

FIGURE 2-5
In the Air Force Space Command (AFSPC), Development Planning—called XR—links the needs of warfighters with the capabilities of the world of science and technology (S&T). NOTE: Operational input to the Space and Missile Systems Center (SMC) will come primarily through Headquarters AFSPC in its role as the SMC's Major Command. For the purposes of this figure, nonmilitary S&T inputs would be grouped in the upper right with the other S&T input providers. SOURCE: Donald E. Wussler, Colonel, Director, Development Planning, Space and Missile Systems Center, USAF. 2010. "SMC/XR Function Brief." Presentation to the committee, April 22, 2010.

the rank—and the perspective and the influence—of some ATC attendees declined: What had at one time been a three-star-level meeting became, in some situations, a conference between colonels, or lieutenant colonels.[35,36]

As with Development Planning, the decline of ATCs represents a significant setback in the pursuit of clear, stable, and realistic requirements. But unlike with the slow recovery of Development Planning, the reinvigoration of ATCs, where necessary, could be done quickly, and the benefits would be felt almost immediately.

[35]Arthur Huber, Colonel, Vice Commander, Aeronautical Systems Center, Wright-Patterson Air Force Base; and Gerald Freisthler, Executive Director, Aeronautical Systems Center, Wright-Patterson Air Force Base. 2010. "Aeronautical Systems Center Involvement in Applied Technology Councils." Presentation to the committee, June 1, 2010.

[36]Ellen Pawlikowski, Major General, Commander, Air Force Research Laboratory (AFRL). Personal communication with the committee, June 11, 2010.

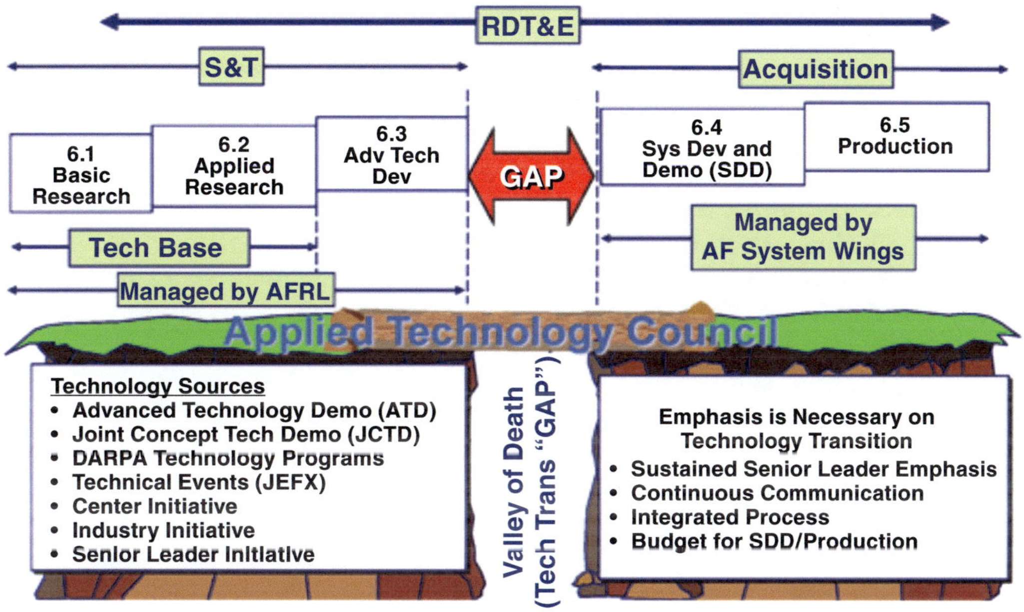

FIGURE 2-6
Applied Technology Councils (ATCs) serve to bridge the Valley of Death technology transition gap between Budget Activities 6.3 and 6.4. SOURCE: Arthur Huber, Colonel, Vice Commander, Wright-Patterson Air Force Base; and Gerald Freisthler, Executive Director, Wright-Patterson Air Force Base. 2010. "Aeronautical Systems Center Involvement in Applied Technology Councils." Presentation to the committee, June 1, 2010.

FINDING 2-3
The decline of Development Planning and, in some quarters, the deterioration in the effectiveness of ATCs have greatly reduced the ability to integrate successfully the interests of warfighters, the S&T community, and acquisition leadership.

Assessing the Maturity of Technology

Technology maturity is a central factor in program risk.[37] Objective measurement of Technology Readiness Level (TRL) assesses a technology's maturity relative to its stage in the development cycle, whereas Manufacturing Readiness Level (MRL) is an analogous measure of manufacturing risk. Integration Readiness Level

[37]The committee defines "technology maturity" using a three-pronged approach, which includes Technology Readiness Level (TRL), Manufacturing Readiness Level (MRL), and Integration Readiness Level.

evaluates the capability of a component or technology to be integrated into a larger system.[38] An accurate appraisal of program risk requires considering all three of these factors, as all will ultimately affect a program's likelihood of success or failure.

Industry has adopted technology development practices that are distinct from those used in the DoD. One particularly crucial feature is that successful technology developers separate technology development from product development. Technology is developed and matured first, and that is followed by the development of a product incorporating the new technology. These steps are not done concurrently.[39] What industry has learned—and the Air Force is seemingly having to *relearn*—is that simultaneously developing new technology within an acquisition program is a recipe for disaster.

To avoid these problems, TRLs are used by industry and government agencies to assess systemic developmental risk by evaluating the maturity of technologies, and then, using that information, to determine readiness to progress from one development phase to the next.

Basic TRL definitions currently used by the DoD for hardware are shown in Figure 2-7. The *Technology Readiness Assessment (TRA) Deskbook* contains more complete descriptions and supporting information for hardware and software TRL definitions, along with descriptions and supporting information.[40] The early application of TRL assessment was quite subjective, but the development of the *TRA Deskbook*[41] makes significant advances toward the application of a uniform and objective assessment process. It will be instructive to observe the extent to which future assessments yield consistent programmatic results.

The requirement for Technology Readiness Assessments is contained in DoD Instruction 5000.02;[42] the Office of the Director, Defense Research and Engineering (DDR&E), plays a key role in technology development and technology maturity assessments. The term "Technology Readiness Level" is often used as a proxy for technology maturity assessments; however, there is a full set of technology maturity metrics that go beyond TRL and include MRL and System Readiness Level (SRL).

---

[38]James Bilbro and Kyle Yang. 2009. "A Comprehensive Overview of Techniques for Measuring System Readiness." *Proceedings of the 12th Annual Systems Engineering Conference, San Diego, Calif., October 26-29, 2009.* Arlington, Va.: National Defense Industrial Association.

[39]Thomas Gehring, Program Manager, 3M Industrial and Transportation Business. 2010. "Technology Development and Innovation at 3M Company." Presentation to the committee, June 7, 2010.

[40]DoD. 2009. *Technology Readiness Assessment (TRA) Deskbook.* Prepared by the Director, Research Directorate, Office of the Director, Defense Research and Engineering (DDR&E). Washington, D.C.: DoD. Available at http://www.dod.mil/ddre/doc/DoD_TRA_July_2009_Read_Version.pdf. Accessed June 10, 2010.

[41]Ibid.

[42]DoD. 2008. Department of Defense Instruction 5000.02. December 8. Available at http://www.dtic.mil/whs/directives/corres/pdf/500002p.pdf. Accessed January 29, 2011.

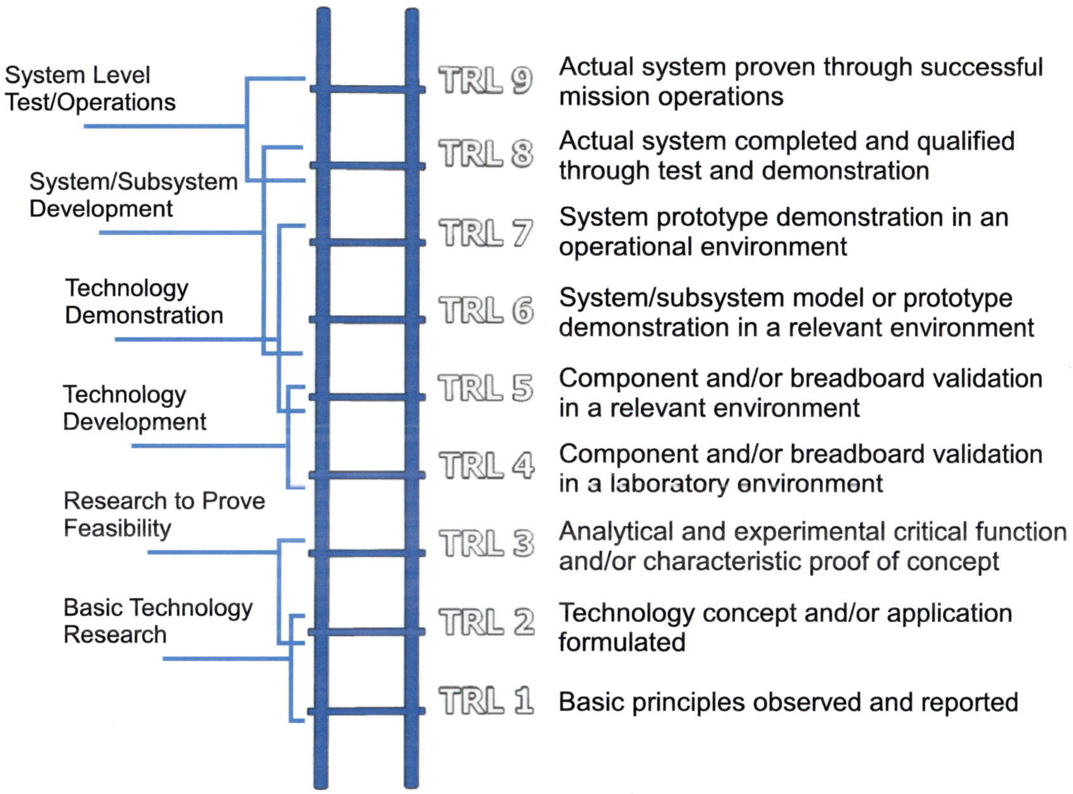

FIGURE 2-7
Department of Defense (DoD) hardware Technology Readiness Levels. SOURCE: Based on information derived from the Department of Defense and the National Aeronautics and Space Administration.

The current DDR&E is placing strong emphasis on Development Planning and prototyping, as well as on the role of systems engineering in the developmental process, to include risk assessment.[43]

The Weapon Systems Acquisition Reform Act of 2009 established several new requirements relating to technological maturity that are summarized below. Among its other provisions, WSARA requires the following:

- Periodic review and assessment of the technological maturity and integration risk of critical technologies of Major Defense Acquisition Programs

[43]Zachary Lemnios, Director, Defense Research and Engineering, Department of Defense. 2010. "Development Planning Initiative Within DoD." Presentation to the committee, May 12, 2010.

(MDAPs), and development of knowledge-based standards against which to measure technological maturity and integration risk;

- An annual report to Congress on technological maturity and integration risk; and
- A report to Congress on additional resources required to implement the legislation.[44]

The first annual DDR&E report[45] to Congress on the technological maturity and integration risk of major DoD acquisition programs was submitted in April 2010. During 2009, DDR&E completed 11 Technology Readiness Assessments of MDAPs and 1 special assessment. The more robust technology readiness oversight role required by the legislation should serve to reinforce the initiatives taken recently by the Air Force to improve the technology maturation process.

A number of Government Accountability Office (GAO) studies in recent years have addressed technology development practices and the importance of technological maturity.[46] A recent article states:

> [A]lthough the Defense Department and the GAO remain at odds over the right technology readiness level (TRL) for new systems, the debate is unlikely to escalate. The Pentagon states that the two organizations continue to disagree on the meaning of mature technology before launching into system development. GAO advocates TRL 7, while the DoD prefers TRL 6. The DoD has taken the position that TRL 6 is adequate at milestone B.[47]

A Case Study on the Importance of Ensuring Technological Readiness: The Joint Strike Fighter

Overly optimistic Technology Readiness Assessments have been a root cause of cost and schedule performance problems on complex programs in the past. The F-35 Joint Strike Fighter (JSF) program, for example, has embraced many new technologies to provide the specified operational performance in a stealthy, multi-role fighter, producible at high production rates. Some notable technologies include a digital "thread" that controls the engineering, tooling, fabrication, assembly, and support systems for the aircraft and also controls several advanced subsystems and components. Although a large majority of the individual technologies incorporated into the F-35 have proven to be sufficiently mature, some of the Engineering and

[44]Weapon Systems Acquisition Reform Act of 2009 (Public Law 111-23, May 22, 2009).

[45]DoD. *Department of Defense Report: Technology Maturity and Integration Risk of Critical Technologies for CY 2009.* Washington, D.C.: DoD.

[46]GAO. 2006. *Best Practices: Stronger Practices Needed to Improve DoD Technology Transition Processes.* GAO-06-883. Washington, D.C.: GAO. Available at http://www.zyn.com/sbir/reference/GAO-d06883.pdf. Accessed June 10, 2010.

[47]Inside Defense News. 2010. "GAO, Pentagon Disagreement on TRLs Unlikely to Escalate." *Inside the Pentagon*, April 22.

Manufacturing Development cost increase has resulted from an unanticipated need for additional technology maturation. Although the increased cost for the EMD program cannot be attributed solely to the shortfalls in TRL, several technologies in retrospect were not at the required TRL 6, and those have contributed to the JSF's delayed development and cost growth.

Public reports indicate that some of the cost increase for the EMD phase of the program has resulted from unanticipated technology maturation during full-scale development of the production configuration.[48] The F-35 EMD program was structured to develop three variants with a high degree of commonality, and cost and schedule were based on assessments of Technology Readiness Levels above TRL 6. For example, one critical technology adopted for the JSF is the electro-hydrostatic actuation system used to power the flight controls. The contractor focused on developing this new technology and demonstrated a prototype subsystem in an F-16 before proposing to use it, but significant problems have been encountered nonetheless. In retrospect, more rigorous maturation of the high-power electronics and the specialized actuators in a representative environment was required for an appropriate level of confidence in the TRL for this complex subsystem.

FINDING 2-4

The absence of independent, rigorous, analytically-based assessments of Technology, Manufacturing, and Integration Readiness Levels will reduce the likelihood of successful program outcomes. Furthermore, despite the existence of clear and compelling examples to the contrary, the Air Force continues to initiate system acquisition prior to completing the required technology development.

Although expert opinions differ about when requirements should be baselined and about the appropriate assessment level to be used as a threshold for entry into EMD, concurrent evolution of technology and requirements should be the norm up to System Requirements Review (SRR.) At SRR, the capabilities of the selected technologies should be clear and the limitations that the technologies place on the operational requirements must be accepted, and either the technology development phase must be extended or the program terminated. One key SRR success criterion to be evaluated is whether the operational requirements can be met given the technology maturation achieved.

FINDING 2-5

After System Requirements Review, stable requirements and a well-defined operational environment are essential to successful technology insertion.

---

[48]More information on the F-35 is available at http://en.wikipedia.org/wiki/Lockheed_Martin_F-35_Lightning_II. Accessed September 2, 2010.

FINDING 2-6
Some important technology insertion efforts have failed to mature due to the lack of (or subsequent loss of) a specific targeted program of record—for example, a new engine technology being developed for a proposed aircraft. Thus, a successful and useful technology may go dormant until a new program can be identified to host it. In this manner, even valuable technology advancements that cannot be inserted in a timely way into a program of record can be relegated to the "Valley of Death."

FINDING 2-7
The array of technology possibilities always exceeds the resources available to pursue them. One result is that the technology planning process tends to overcommit available resources and does not always ensure that every technology investment has an executable plan (with a corresponding budget) that enables near-term production readiness.

Resources

Stable, clear, feasible and well-understood requirements are essential to the success of acquisition programs. Equally important are stable funding and robust processes that can reliably create satisfactory programmatic outcomes. As seen above, some of these processes are problematic. Some current processes are inadequately implemented, and others—like the ATCs—work for a time and then slide into disuse. Other processes do work—like the TRL system successfully used in industry and elsewhere in government (NASA, for example)—but for a variety of reasons fail to be used in a disciplined way, with risky and insufficiently proven technology comprising important parts of major programs. But the most significant—and surprising—process shortfall was the lack of an articulate and formal Air Force-level S&T strategy. To the contrary, a number of Air Force stakeholders asserted that there is no such overarching strategy, often unfavorably comparing the Air Force's failure in this area to what they considered the more successful Future Naval Capabilities process of the U.S. Navy.[49,50]

Even a successfully resuscitated DP capability is not a substitute for an Air Force-level S&T strategy. The developmental planners at each Product Center strive to identify and prioritize technology development activities to match the require-

[49]Neil Kacena, Vice President, Advanced Development Programs Deputy, Lockheed Martin Aeronautics Company. 2010. "Technology Development: Approaches and Challenges." Presentation to the committee, May 12, 2010.

[50]A. Thomas Young, Retired Executive Vice President, Lockheed Martin Corporation. 2010. "Best Practices." Presentation to the committee, May 12, 2010.

ments of their particular Major Command (MAJCOM) customers; nevertheless, the committee heard from multiple presenters that the Air Force does not attempt in any disciplined way to set technology priorities across the entire service.

A telling example of this need for an Air Force-level technology prioritization strategy can be found in the *Technology Horizons* study recently conducted by the Air Force Chief Scientist.[51] *Technology Horizons* is the most recent in a succession of major S&T vision studies conducted at the Headquarters Air Force level. The study is an effort long overdue to help define key, priority S&T investments to provide the Air Force with the capabilities that it will need over the next 10 to 20 years. However, the study is focused on and written from the perspective of the S&T world. Although it identifies potential capability areas that might benefit from Air Force S&T activities, it does not answer the operationally oriented question of what future capabilities the Air Force needs to acquire. In other words, the technology opportunities described in the study need to be matched to the requirements established by operational Air Force organizations in order to optimize the Air Force's S&T investments.

Similarly, ongoing efforts to reinvigorate ATCs that have fallen into decline will not substitute for an Air Force-level technology strategy. Balancing modernization needs and existing program support with available resources is a constant challenge. Pressures from oversubscribed Air Force budgets repeatedly drive short-suspense reprogramming actions on research and development funding, often with little in the way of rational analysis.[52] Absent a technology strategy and prioritized list of technology maturation needs, the Air Force POM and budget process will not provide a solid foundation for future acquisition program success, as illustrated in Figure 2-8.

FINDING 2-8

The Air Force lacks an effective process for determining which technology transitions to fund.

Nonetheless, a revived and rechartered ATC process could provide a forum for integrating MAJCOM capability needs with technology opportunities and technology maturation funding priorities. Consequently, the Air Force might consider means to link ATCs with MAJCOM representation to an Air Force-level S&T council, as shown in Figure 2-9, that provides top leadership consideration of all

---

[51]Air Force Chief Scientist. 2010. *Report on Technology Horizons: A Vision for Air Force Science and Technology During 2010-2030.* Washington, D.C.: Department of Defense. Available at http://www.airforce-magazine.com/SiteCollectionDocuments/TheDocumentFile/Strategy%20and%20Concepts/TechnologyHorizonsVol1_2010.pdf. Accessed August 26, 2010.

[52]Donald Hoffman, General, Commander, Air Force Materiel Command, USAF. Personal communication to the committee, July 15, 2010.

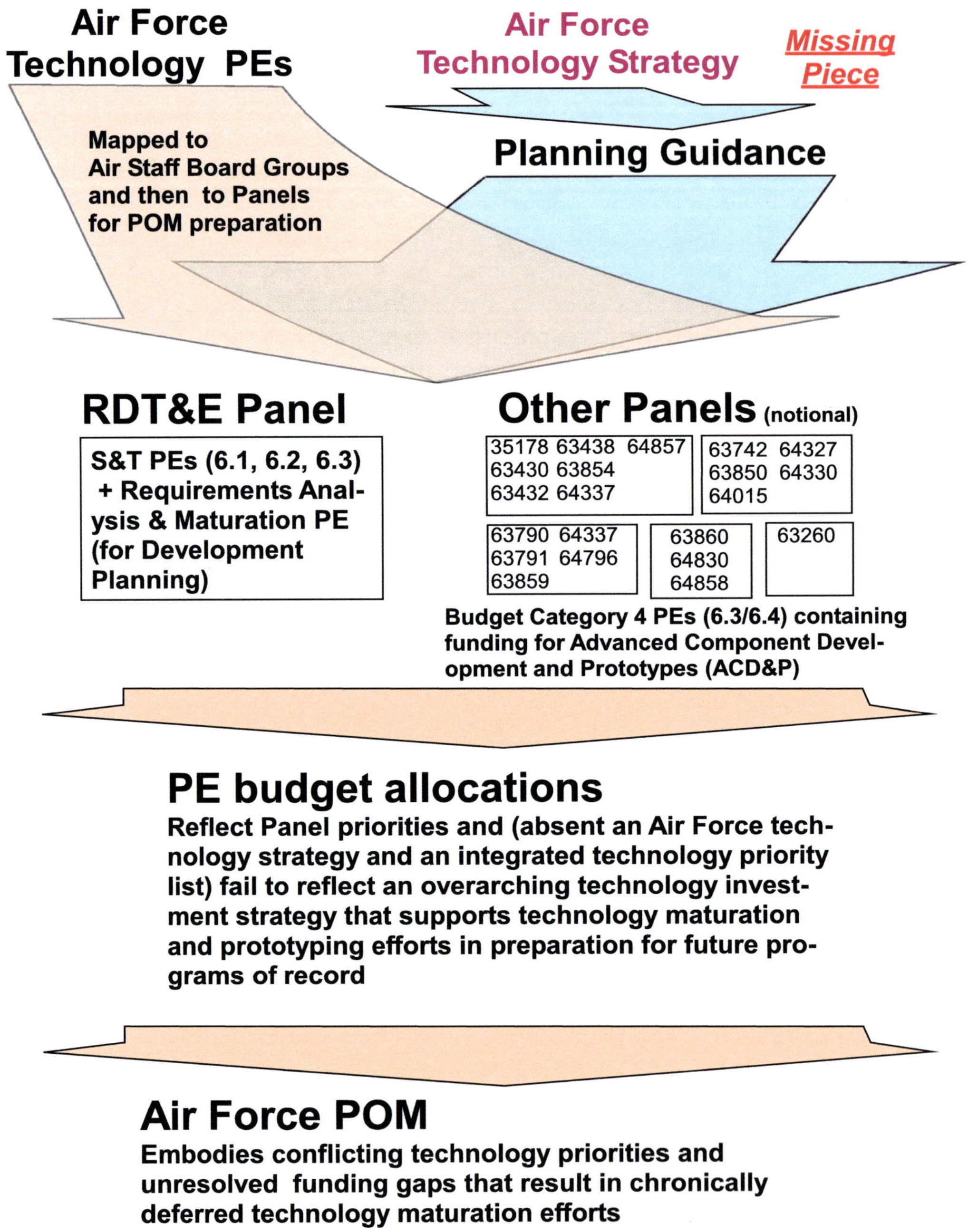

FIGURE 2-8
Air Force budget process. Current technology development funding is spread across multiple budget panels without an overarching investment strategy or prioritization. This could result in technology gaps in multiple acquisition programs.

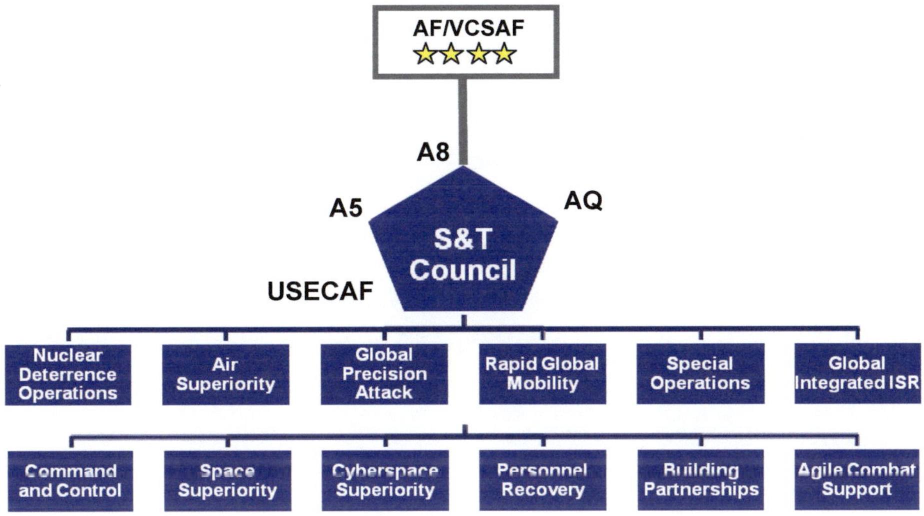

FIGURE 2-9
A possible Air Force-level science and technology (S&T) council organization structure under consideration by an Air Force Tiger Team chartered to examine S&T strategy governance and strategic planning processes. The S&T council would potentially review and approve all S&T guidance and oversee technology transition progress. SOURCE: Michael Kuliasha, Chief Technologist, Air Force Research Laboratory. 2010. "AFRL Perspective on Improving Technology Development and Transition." Presentation to the committee, May 13, 2010.

MAJCOM priorities, all laboratory S&T contributions, and *all* appropriate (6.3 and 6.4) funding. Such a process is needed if the Air Force is ever to have a strategic technology planning process.

Air Force leadership, after watching the number of funded Advanced Technology Demonstrations dwindle from 65 in 2000 to just 2 in 2009, recently chartered a Tiger Team to examine options for strengthening the S&T strategy planning process.[53] The Tiger Team will identify opportunities for improvement in communication and governance that can lead to consistent S&T and transition priorities across all organization levels and to improved visibility and accountability of S&T needs and solutions. Tiger Team members are drawn from organizations across the Air Force and DoD and are assessing a range of possible S&T strategy governance options.

[53]Ibid.

FINDING 2-9
Although the Air Force Chief Scientist has developed an "art of the possible" science and technology strategic plan for the 2010 to 2030 time frame, there exists no Air Force-level unifying strategy, inextricably linked to operational requirements, to guide decision making for science and technology investments.

FINDING 2-10
Successful technology development and technology transition require (1) integration of warfighter requirements with science and technology investments and systems acquisition strategies, and (2) close collaboration among all government and industry partners.

FINDING 2-11
MAJCOM ownership of Budget Category 4 Program Elements and the current Air Force Budget formulation process do not provide development planners with sufficient priorities for execution of maturation funding. At a higher level, the Air Force lacks an overarching strategy for technology development, or a process that involves key decision makers. As a result, there is no integrated view of warfighter needs and technological possibilities, and there is inadequate guidance for determining what technology transitions to fund.[54]

The Right People

The third of the "Three Rs"—the right people—is the most important. Without the right people, programs are more likely to fail, even when requirements and resources are addressed successfully. The phrase "right people" implies that there are enough people with the necessary knowledge and experience, in both government and industry, who are educated, trained, mentored, experienced, credible, empowered, and trusted to do the job at hand—that is, people who can, with the resources, meet the requirements and deliver needed capability to the warfighter.

---

[54]Budget Activity 4, Advanced Component Development and Prototypes (ACD&P): Efforts necessary to evaluate integrated technologies, representative modes, or prototype systems in a high-fidelity and realistic operating environment are funded in this budget activity. The ACD&P phase includes system-specific efforts that help expedite technology transition from the laboratory to operational use. Emphasis is on proving component and subsystem maturity prior to integration in major and complex systems, and may involve risk-reduction initiatives. Program elements in this category involve efforts prior to Milestone B and are referred to as advanced component development activities and include technology demonstration. Completion of Technology Readiness Levels 6 and 7 should be achieved for major programs. Program control is exercised at the program and project level. A logical progression of program phases and development and/or production funding must be evident in the FYDP. *DoD Financial Regulation*, Volume 2B, Chapter 5, June 2004.

Losses suffered by the Air Force acquisition workforce over the past two decades have been significant. Highlighted in the Kaminski report as well as in other reports were the ramifications of mandated reductions in acquisition personnel in the 1990s.[55] Further, the Air Force Acquisition Improvement Plan states:

> The Air Force acquisition workforce is staffed with outstanding men and women dedicated to their mission and their country. . . . However, while they perform top quality work, we have failed to adequately manage their professional development and maintain sufficient numbers of these experienced professionals. The result is an acquisition workforce eager and willing to take on any challenge, but in many cases one that is inadequately prepared for the task at hand. In some cases, the workforce lacks the necessary training or education to accomplish the mission. In others, the workforce simply does not have the depth of experience or specific skill sets necessary to accomplish the critical tasks.
>
> As we better develop our workforce, we must also ensure it is appropriately sized to perform essential, inherently governmental functions and is flexible enough to meet continuously evolving demands. The size of the Air Force acquisition workforce, as currently defined, was decreased from a total of 43,100 in 1989 to approximately 25,000 in 2001 where it has remained since.[56]

The cumulative impact of all of the reductions and changes to the workforce can best be summarized in the following statements from a 2009 report of Business Executives for National Security (BENS):

> Today the government too often finds itself with minimally experienced and transient individuals leading major acquisition programs, able to attract new people only after long delays, unable to couple rewards to performance, and with many senior positions simply unoccupied. Talented and dedicated people can often overcome a poor organizational structure, but a good organizational structure cannot overcome inadequate performance. When qualified people are combined with sound organizations and practices, success is virtually assured. The acquisition process, unlike most government pursuits, is a business function. It demands skills and talents that are far more common to the business world than to government and military operations.[57]

In building Lockheed Martin's Skunk Works, Kelly Johnson learned the importance of having good people and that quantity was no substitute for quality and

[55]NRC. 2008. *Pre-Milestone A and Early-Phase Systems Engineering: A Retrospective Review and Benefits for Future Air Force Systems Acquisition.* Washington, D.C.: The National Academies Press.

[56]USAF. 2009. *Acquisition Improvement Plan.* Washington, D.C.: Headquarters USAF. May 4, p. 4. Available at http://www.dodbuzz.com/wp-content/uploads/2009/05/acquisition-improvement-plan-4-may-09.pdf. Accessed June 11, 2010.

[57]Business Executives for National Security. 2009. *Getting to Best: Reforming the Defense Acquisition Enterprise. A Business Imperative for Change from the Task Force on Defense Acquisition Law and Oversight,* p. 7. Available at http://www.bens.org/mis_support/Reforming%20the%20Defense.pdf. Accessed June 10, 2010.

experience. To paraphrase Johnson: "You can't stack enough average people high enough to equal one good person." But that is exactly the situation facing the Air Force today. The loss of quality and experience over the past 20 years means that with few experienced people left to mentor newer hires, the Air Force must rely on large numbers of inexperienced and unproven acquisition professionals. One presenter to the committee spoke of a program to which the contractor had assigned 80 engineers, who stood stunned as a government review team arrived with 137 participants, most of them junior military and civilian employees.[58] When the number of "checkers" nearly doubles the number of "doers," it is hard to see that as a path to recapturing acquisition excellence.

Strong and innovative hiring efforts are under way and are aimed not only at encouraging new entrants to join the workforce, but also at capturing mid-career professionals from other agencies and industries. Those efforts are necessary and will pay off years down the road, but right now and for the foreseeable future, the Air Force is learning a hard lesson, similar to the lesson from the demise of Development Planning: An asset can be lost in the blink of an eye, but rebuilding it is the work of decades.

FINDING 2-12

The size and experience of the Air Force technology and development planning workforces are inadequate. Despite concerted efforts to fulfill the vision of a revitalized Development Planning function, recovery in this area will take a substantial period of time and a constancy and consistency of purpose from Air Force leadership.

CONCLUDING THOUGHTS

Historically, successful acquisition programs have followed a dedicated period of technology development and maturation. In the late 1990s, Congress eliminated funding for key organizations and processes that enabled that technology development and maturation (e.g., the Product Center Development Planning Organizations, known as XRs). This resulted in the dispersal of the DP workforce, as resources were reassigned to other activities. The resultant technology development and maturation vacuum was, to some extent, filled by aerospace industry firms, advisory and assistance support contractors, and other ad hoc efforts, many of which lacked the focus and coherence of previous DP organizations and processes. Product Centers recognized the risk to program success caused by this situation

[58]Dwyer Dennis, Brigadier General, Director, Intelligence and Requirements Directorate, Headquarters Air Force Materiel Command, Wright-Patterson Air Force Base, Ohio. 2010. "Development Planning." Presentation to the committee, March 31, 2010.

and began homegrown efforts to restore the XRs, but the resultant organizations have remained chronically underfunded, understaffed, and underequipped.[59,60,61] Other efforts, such as the ATCs, at one time fostered timely and effective decision making regarding scarce technology maturation and funding. However, in some areas, ATCs and similar initiatives have been allowed to wither.

Meanwhile, poorly performing and failed programs have caused great frustration in the Congress and the OSD, leading to a serious erosion of trust of the Air Force's stewardship of force modernization efforts. This distrust has resulted in statute- and policy-driven increases in program oversight during all phases of the acquisition cycle. This increased oversight is moving earlier in the process, being applied to preacquisition technology development activities (e.g., Material Development Decisions to Milestone B). One result is an increase in the number of "checkers" at the expense of the "doers"—an overemphasis on people performing review and oversight rather than executing the basics of technology development and program management. The "right people" means the right numbers of people, with the right experience and skills, doing the right things.

Increased oversight also has led, at times, to unrealistic program goals prior to Milestone B. Recently passed legislation and resultant DoD policy initiatives—for example, Section 852 of the 2008 National Defense Authorization Act (NDAA), competitive prototyping, DoD Instruction 5000.02, AIP, and WSARA—appear to address some of the negative impacts of the dissolution of DP organization and processes. However, sufficient funding levels are not yet evident, and the growing oversight environment, particularly pre-Milestone B, does not bode well for the full restoration of a robust preacquisition technology development capability.

[59]Donald E. Wussler, Colonel, Director, Development Planning, Space and Missile Systems Center, USAF. "SMC/XR Function Brief." Presentation to the committee, April 22, 2010.

[60]Charles Kelley, Director, Capability Integration, Electronic Systems Center, Hanscom Air Force Base, Massachusetts. "ESC/XR Function Brief." Presentation to the committee, April 21, 2010.

[61]Edward Stanhouse, Colonel, Requirements and Capabilities Integration, Aeronautical Systems Center. "ASC/XR Function Brief." Presentation to the committee, April 22, 2010.

3

Government and Industry Best Practices

The objective of this chapter is to report on and discuss industry and government best practices addressing both evolutionary (deliberate) and revolutionary (rapid) technology development in areas correlated to the findings presented in Chapter 2. Case studies are discussed below, highlighting programs and procedures that have achieved positive outcomes for customers and users. Although every program and process could be further evaluated for failure modes and lessons learned that might be applied to future endeavors, the selected examples instead showcase the elements that clearly illustrate best practices.

BEST PRACTICES

Although creating a perfect technology development process for the United States Air Force may be too lofty a goal, the desired end state is enabling the "Three Rs"—(1) Requirements, (2) Resources, and (3) the Right People—that give process stakeholders the clear path that they need to work together successfully. Under the umbrella of these "Three Rs" are several critical factors, organized below as best practices. The Department of Defense (DoD) acquisition process is an intricate web of policies, organizations, processes, people, and priorities. Only when these forces are operating harmoniously can the process meet warfighter needs efficiently and effectively.

EXAMPLES OF GOVERNMENT BEST PRACTICES

The complexity of the DoD acquisition process can render the process slow and expensive. However, the work flow can be tailored to address situational requirements, as permitted by DoD Instruction 5000.02. In fact, the document's outlined purpose includes a direct reference to the importance of a tailored approach, authorizing "Milestone Decision Authorities (MDAs) to tailor the regulatory information requirements and acquisition process procedures in this Instruction to achieve cost, schedule, and performance goals."[1] The instruction goes on to recommend areas in which a customized approach can be a critical success factor. In short, the DoD itself now recommends adapting the acquisition process, and success stories have arisen directly from this customized approach.

Joint Improvised Explosive Device Defeat Organization

Improvised explosive devices (IEDs) became an everyday danger to troops deployed in Iraq during 2003, with some DoD estimates of 20 attacks daily even during those early days of Operation Iraqi Freedom.[2] This highly lethal and adaptive threat necessitated a more flexible acquisition model to neutralize an enemy with rapidly evolving tactics, techniques, and procedures. As the United States sought to counter this threat using add-on armor plates and other forms of protection on ground vehicles, insurgents became more creative in their emplacement methods as well as their technologies. For example, IEDs emerged with explosively formed penetrators (EFPs), which focus the explosive energy and projectile in a specific direction.[3] The immediate need for more robust, powerful IED protection took center stage.

The Joint Improvised Explosive Device Defeat Organization (JIEDDO) was formally established by the Office of the Secretary of Defense (OSD) in 2006, having grown from the Army's task force formed in 2003 to counter the IED threat in Iraq and Afghanistan.[4] The organization's stated mission is "to rapidly provide Counter

[1]DoD. 2008. Department of Defense Instruction 5000.02. December 8. Available at http://www.dtic.mil/whs/directives/corres/pdf/500002p.pdf. Accessed on January 29, 2011.

[2]Donald P. Wright, Timothy R. Reese, and the Contemporary Operations Study Team. 2008. *On Point II: Transition to the New Campaign, The United States Army in Operation Iraqi Freedom May 2003-January 2005.* Fort Leavenworth, Tex.: Combat Studies Institute Press. Available at http://www.globalsecurity.org/military/library/report/2008/onpoint/index.html. Accessed August 6, 2010.

[3]Information on an explosively formed penetrator warhead is available at http://www.globalsecurity.org/military/systems/munitions/bullets2-shaped-charge.htm. Accessed August 6, 2010.

[4]More information on the Joint Improvised Explosive Device Defeat Organization is available at https://www.jieddo.dod.mil/about.aspx. Accessed September 2, 2010.

Improvised Explosive Device capabilities in support of the Combatant Commanders and to enable the defeat of the IED as a weapon of strategic influence."[5]

Part of JIEDDO's success lies in its acquisition model, which is built on the premise that the organization must be as quick and agile as the enemy is. In this case, the enemy makes broad use of commercial, off-the-shelf technologies to rapidly bombard warfighters and civilians with new threats (e.g., IEDs triggered by cordless telephones and garage-door openers). Combining this urgent need with congressional relief on funding constraints enabled timely expenditures on research and development, procurement, and operations and maintenance, without the limitations commonly found in more routine technology development and acquisition. JIEDDO's tailored procurement process allowed it to respond to joint urgent operational needs (thus bypassing the lengthy Joint Capability Integration and Development System [JCIDS] model), which significantly reduced response time.

JIEDDO also formalized its acquisition model, called the Joint IED Defeat Capability Approval and Acquisition Management Process (JCAAMP), in 2007. Shown in Figure 3-1, the JCAAMP is focused on rapidly delivering new capabilities to the field to defeat emergent technologies like the EFP.[6] In addition, JIEDDO outlined a three-pronged approach to focus its efforts—attack the network, train the force, and defeat the device. JCAAMP characteristics include the following: support for flexible points of entry depending on maturity, effective testing and evaluation requirements, continued initiative assessment after initial delivery to warfighters, continuous enforcement of risk reduction, and formal oversight of acquisition during every phase.

Effective technology does not always require high levels of complexity. In 2006, JIEDDO zeroed in on the trigger mechanism for the EFP to reduce its accuracy, using a simple technological solution called Rhino, made of vehicle glow plugs and batteries housed in a munitions can. When extended in front of the lead vehicle in a convoy, this device causes passive infrared-triggered IEDs to detonate prior to the vehicle's entering the kill zone. With the U.S. Army's support, JIEDDO tested this capability in the continental United States, then competed the solution for production. By the end of 2008, more than 16,000 Rhino II systems had been produced and deployed under JIEDDO's JCAAMP funding. Technology enhancements were delivered to the field later through upgrade kits, which included safety and performance improvements. Additional technological upgrades were made to adapt the system for the unique terrain requirements of Afghanistan (Rhino III), and

[5]The mission statement for the Joint Improvised Explosive Device Defeat Organization is available at https://www.jieddo.dod.mil/index.aspx. Accessed August 6, 2010.

[6]DoD. 2009. *Joint Improvised Explosive Device Defeat (JIEDD) Capability Approval and Acquisition Management Process.* Joint Improvised Explosive Device Defeat Organization Instruction 5000.01. Washington, D.C.: Department of Defense. Available at https://www.jieddo.dod.mil/content/docs/20091106_JCAAMP_update.pdf. Accessed August 6, 2010.

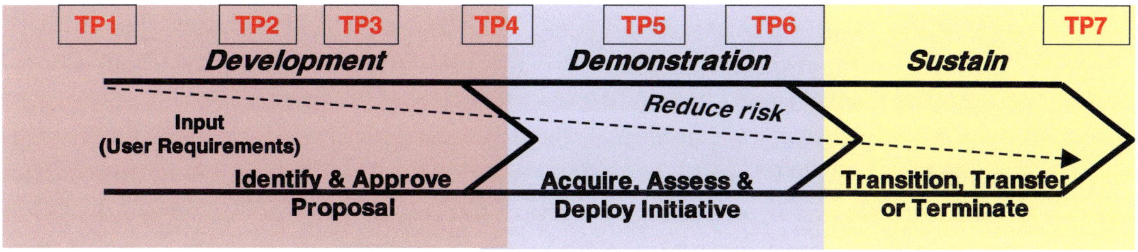

FIGURE 3-1
The Joint Improvised Explosive Device Defeat Organization's (JIEDDO's) Capability Approval and Acquisition Management Process acquisition model. NOTE: *Pre-TP 1:* Potential solutions to published operational or technology capability gaps received by way of the BAA Information Delivery System are adjusted for feasibility. S&T or proofs of concept supporting requirements are considered development efforts and are matured until eligible to proceed to successive TPs. *TP 1:* Gain approval from the vice director for the proposal to proceed. Complete evaluation of technology merit against operational capability gaps and readiness levels. *TP 2:* LOO Portfolio Manager approval to proceed to the JIEDDO Requirements, Resources, and Acquisition Board (JR2AB). *TP 3:* JR2AB recommendation to proceed to Joint Improvised Explosive Device Defeat (JIEDD) Integrated Process Team (JIPT). *TP 4:* JIPT endorsement and Decision Memorandum signed. *TP 5:* Decision by vice director to deploy an initiative for employment by operational units. *TP 6:* Decision to field a proven initiative. Decision to transition, transfer, or terminate (T3) initiative, as appropriate. *TP 7:* Complete transition or transfer of initiatives to services or agencies; or terminate. SOURCE: Department of Defense. 2009. *Joint Improvised Explosive Device Defeat (JIEDD) Capability Approval and Acquisition Management Process.* Joint Improvised Explosive Device Defeat Organization Instruction 5000.01. Washington, D.C.: Department of Defense. Available at https://www.jieddo.dod.mil/content/docs/20091106_JCAAMP_update.pdf. Accessed August 6, 2010.

Rhino was transitioned into a program of record.[7] As demonstrated by JIEDDO's successes against a difficult and tightly scoped mission, an institutionalized focus on the flexible, efficient acquisition of technology is critical to achieving results. JIEDDO's tailored acquisition process allows the right people to provide the right solution at the right time, thus quickly responding to deadly threats.

Capabilities Development for Rapid Transition

As with the JIEDDO example, the U.S. Army's Capabilities Development for Rapid Transition (CDRT) method is designed to get "capabilities to the Soldier quicker and with less fiscal and schedule risk than through the standard acquisition process."[8] The CDRT process identifies high-potential technologies being used

[7]Joint Improvised Explosive Device Defeat Organization policy on "Defeat the Device" is available at https://www.jieddo.dod.mil/defeat.aspx. Accessed August 6, 2010.

[8]U.S. Army. 2008. *Spiral Technology and Capabilities Development for Rapid Transition to the Army.* 2008 U.S. Army Posture Statement. Available at http://www.army.mil/aps/08/information_papers/transform/Spiral_Technology_and_Capabilities.html. Accessed August 6, 2010.

successfully in theater on a small scale, then assesses their broader applicability. The most promising are selected for recommendation to the Army, either as new programs of record or as accelerations into existing programs.[9]

With its emphasis on analyzing battlefield trends in order to identify emerging needs, the CRDT is one of three primary technology development approaches used by the Army for technology insertion and technology development. Another approach is in response to an operational needs assessment; through this approach, commanders in the field can identify specific requirements that are then verified and responded to by the Army. Finally, the more traditional approach of ongoing technology development is focused on continuous improvement of warfighter capabilities.[10] As expressed by the Army:

> The primary weakness of the current acquisition process is that it can take up to seven years to field a required capability. . . . CDRT significantly reduces the time required to get new technologies into the field, while ensuring critical capabilities are fully documented and supported over time.[11]

Thus, the CDRT is a critical element in the Army's multipronged approach to procurement, which provides multiple options while working within the established organizational structure. Because the approach embraces customization from the onset, the Army can drive innovation at the appropriate pace.

Big Safari

Big Safari, an Air Force rapid-acquisition program in existence since the 1950s, is responsible for the rapid acquisition, modification, testing, fielding, and sustainment of selected capabilities as determined by the Assistant Secretary of the Air Force (Acquisition). In 2002, Big Safari took its own steps to rapidly transition critical unmanned aircraft intelligence, surveillance, and reconnaissance capability to warfighters by providing direct sensor data to troops on the ground.[12] Big Safari's success is directly tied to the ability to integrate off-the-shelf technologies in support of urgent operational needs.

---

[9]Daniel Wolfe, CEO, Universal Solutions International, Inc. 2010. "Remarks to the National Research Council Air Force Preacquisition Technology Development Study Panel." Presentation to the committee, June 7, 2010.

[10]U.S. Army. 2008. *Spiral Technology and Capabilities Development for Rapid Transition to the Army.* 2008 U.S. Army Posture Statement. Available at http://www.army.mil/aps/08/information_papers/transform/Spiral_Technology_and_Capabilities.html. Accessed August 6, 2010.

[11]Ibid.

[12]Chris Pocock. 2008. "L-3 Shows Latest, Handheld ROVER Terminal." *Aviation International News*, July 14. Available at http://www.ainonline.com/news/single-news-page/article/l-3-shows-latest-handheld-rover-terminal-16649/. Accessed August 6, 2010.

For example, in January 2002, the Big Safari program director was briefed on a concept that would provide ground troops the capability to receive video feeds from Predator unmanned aircraft in flight. By October, the Remotely Operated Video Enhanced Receiver (ROVER) working group was formed, including members of the U.S. Special Operations Command, the Army's Special Forces Command, and other government agencies.

In order to meet full system specifications, Special Operations Tactical Video System (SOTVS) transmitters and receivers would have had to be installed on the Predator aircraft. Instead, the Big Safari team traded the fully integrated solution for an 80 percent capability. The team decided that decoupling a receiver from an aircraft would be the quickest way to deploy the system. The modified system with proven Technology Readiness Level (TRL) 7 maturity quickly established a one-way link from Predator aircraft to ground units. In fact, this decision enabled Big Safari to deliver a successful prototype of the ROVER system to the C Company, 3rd Special Forces Group (Airborne), within just 2 weeks.[13]

Today, this capability is present across the Army in L-3 Communications' ROVER III, IV, V, and VI systems, as well as AAI Corporation's One System® Remote Video Terminal and Sierra Nevada Corporation's (SNC's) Tacticomp™ system. In this case, the existence of an organization dedicated specifically to rapid technology development, and its key decision to incorporate proven and easy-to-integrate technologies to achieve an "80 percent solution," enabled the team to meet its goals.

FINDING 3-1
Tailored processes can enable rapid technology insertion.

Future Naval Capabilities Process

The United States Navy (USN), including the Marine Corps, adopted the Future Naval Capabilities (FNC) process in 1999, shifting its investment focus from individual technology goals to the most vital future capabilities that can be fielded in 3 to 5 years. This forces the Navy's near-term science and technology (S&T) efforts to center on delivering maturing technology to acquisition managers for timely incorporation into platforms, weapons, and sensors.[14]

[13]"The Down Side of a Hack." January 2010. Available at http://www.strategypage.com/htmw/htecm/articles/20100110.aspx. Accessed August 6, 2010.

[14]Office of Naval Research (ONR). 2008. *ONR Selects 12 to Share $1.2 Million Research Funding Recipients Awarded Seed Money for Specific Promising and Innovative S&T Ideas.* Office of Naval Research Press Release. Available at http://www.onr.navy.mil/en/Media-Center/Press-Releases/2008/12-Share-Funding-Award.aspx. Accessed August 6, 2010.

FNC programs are administered through the Office of Naval Research (ONR), charged in the National Defense Authorization Act of 2001 (Public Law 106-398) with managing "the Navy's basic, applied and advanced research to foster transition from science and technology to higher levels of research, development, test and evaluation."[15] This structure enables the ONR to fund and manage all phases of S&T development (6.1 through 6.4), and as a result, its experts can see programs through from beginning to end. The ONR executes FNC programs through a vast network of expertise, including hundreds of industry partners as well as numerous universities and nonprofit organizations.[16] FNC activities account for roughly 30 percent of the Navy's S&T budget, or approximately $500 million per year.[17]

The FNC program underwent restructuring in 2005 to align with leadership's strategy for enabling capabilities that could be delivered within a 3- to 5-year period, closing known gaps experienced by the warfighter. To accomplish this objective, ONR bundles discrete but interrelated S&T products. Performance and maturity must be quantifiable and meet pre-negotiated exit criteria. Integrated Product Teams (IPTs) are composed of senior leaders, acquisition professionals, and technical personnel who are charged with directing technology development activities.[18]

The IPTs are overseen by the S&T Corporate Board, which includes the Vice Chief of Naval Operations, the Assistant Secretary of the Navy, and the Assistant Commandant of the Marine Corps. These four-star-level leaders ensure that the most important DoD and naval strategies are reflected in naval S&T priorities.[19]

This group is supported by the Technology Oversight Group, composed of several two-star Navy and Marine Corps leaders. The group is charged with oversight, integration, and investment decisions across FNC programs, as well as with approving individual FNC programs if they are found to close critical S&T gaps, rather than simply adding a new product into the mix.[20] In addition, a Require-

[15]Congressional Record, Senate. June 14, 2000. National Defense Authorization Act for Fiscal Year 2001, Amendment No. 3382. Congressional Record Online via GPO Access. Available at http://frwebgate2.access.gpo.gov/cgi-bin/TEXTgate.cgi?WAISdocID=Z7Q17N/2/1/0&WAISaction=retrieve.

[16]ONR. 2009. *Innovation Newsletter.* Arlington, Va.: ONR. Available at http://www.onr.navy.mil/Science-Technology/Directorates/office-innovation/~/media/AE5C1A7063244DFDABF807353421F3D0.ashx. Accessed August 6, 2010.

[17]USN. 2003. *Naval Transformation Roadmap.* Available at http://www.navy.mil/navydata/transformation/trans-pg92.html. Accessed September 2, 2010.

[18]NRC. 2004. *2003 Assessment of the Office of Naval Research's Marine Corps Science and Technology Program.* Washington, D.C.: The National Academies Press.

[19]Ibid.

[20]More information on FNC programs is available on the Navy Web site at http://www.navy.mil/navydata/transformation/trans-pg92.html. Accessed August 9, 2010.

ments IPT, composed of two-star members from several functional disciplines, is responsible for transition resource programming, the preparation of required acquisition documentation, the prioritizing of new start proposals, and the coordination of sea trials.[21] As illustrated in Figure 3-2, this robust organizational structure of checks and balances ensures that every program decision is built around cohesion with strategic needs, coordination of efforts, and emphasis on streamlined decision making.

Transition plans are required before an FNC program can be approved, but there exists the flexibility to progress if a project with high transformation potential is brought forward without a clear transitional picture.[22] Once approved, TRL 6 maturity is expected within 5 years. Technology Transition Agreements are put in place before a program of record can be established or a specification can be written. These contracts, co-signed by the program manager, the developer, and the military sponsor, describe the FNC products, the level of technical risk, the TRL, the exit criteria, and the transition schedule, enabling the necessary budgeting for program transition and minimizing the risk that valuable technology will fail to mature and be applied (i.e., fall into the Valley of Death).

More so than in the Air Force, ONR S&T managers are held accountable and rewarded for the successful transition of programs from FNCs to full-fledged programs of record. This maintains focus and the motivation to ensure that the fleet's requirements are understood, to develop enabling capabilities to address any gaps, and to reach the required goal of TRL 6 within 5 years. By way of example, all senior government employees with oversight or direct responsibility for an FNC are required, at their annual review, to declare the percentage of their programs that have transitioned. Their resulting bonuses are related to this success measure.

By strategically managing funding for all phases of S&T development, incorporating the right people at every stage of the process, and charting the course clearly with established time frames, milestones, requirements, and ties to warfighter needs, the Navy has created a culture of accountability and performance with its FNC process.

---

[21] Available at http://www.onr.navy.mil/en/Science-Technology/Directorates/Transition/Future-Naval-Capabilities-FNC.aspx. Accessed August 9, 2010.

[22] See http://www.onr.navy.mil/en/Science-Technology/Directorates/Transition/Future-Naval-Capabilities-FNC.aspx. Accessed August 9, 2010.

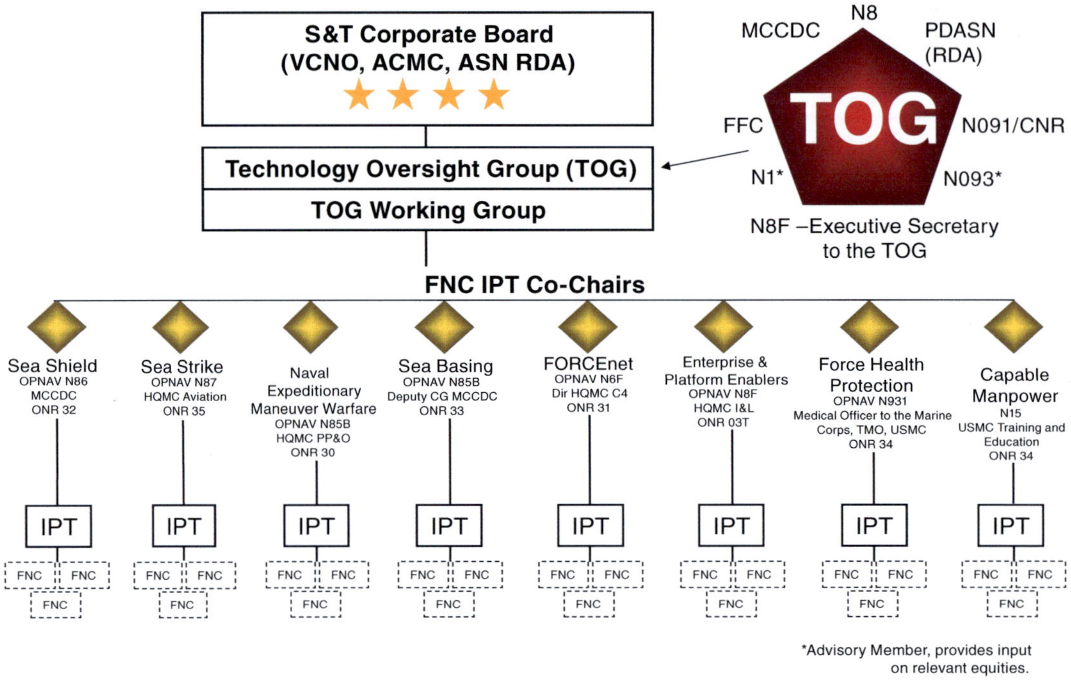

FIGURE 3-2
The organizational structure of Future Naval Capabilities (FNC). SOURCE: Lawrence Schuette, Director, Office of Innovation, Office of Naval Research.

Army S&T Objective Process

The Army also leverages a rigorous process to identify, select, and pursue S&T investments, which generally fall into the categories of basic research, applied research, and advanced technology development.[23,24,25] Army Technology Offices (ATOs) are established on the basis of the areas that have been identified as critical

[23]Mary Miller. 2009. *Technology Transition—Lessons Learned.* Report produced by the Director for Technology, Office of Deputy Assistant Secretary for Research and Technology/Chief Scientist. Washington, D.C.: Department of the Army.

[24]Department of the Army. 2007. *Army Science and Technology Master Plan: Charting the Future of S&T for the Soldier.* Report by the Office of Deputy Assistant Secretary of the Army for Research and Technology. Washington, D.C.: Department of the Army. Available at http://www.carlisle.army.mil/dime/documents/JPLD_AY08_Lsn%207_Reading%204_ASTMP.pdf. Accessed September 2, 2010.

[25]Thomas H. Killion, Deputy Assistant Secretary of the Army for Research and Technology/Chief Scientist. 2010. *Army Science & Technology.* Slides provided at the National Defense Industrial Association 11th Annual Science and Engineering Technology Conference, Charleston, S.C., April 13, 2010. Available at http://www.dtic.mil/ndia/2010SET/Killion.pdf. Accessed September 2, 2010.

to outfitting the future force. There are three types of ATOs, each representing a unique S&T development process:

- *ATO-Demonstration (ATO-D)*—a program focused on transitioning a specific technology to the warfighter within 2 to 4 years.
- *ATO-Research (ATO-R)*—a research-focused activity intended to develop an immature technology further, possibly as a leading step toward a future ATO-D program.
- *ATO-Manufacturing Technology (ATO-M)*—a program that centers on reducing cost through improved strategies for production.[26,27]

Although the three categories of ATO have differing scopes based on program goals and potential barriers to success, all three share a rigidly enforced structure. ATOs must have prescribed milestones and schedules, metrics that are agreed on by all participating parties, and established expectations for technology maturity by the program's conclusion.[28] Figure 3-3 illustrates that ATOs typically require a TRL of 6 or higher; this decreases risk during technology transition.[29]

Funding is only provided to those ATOs for which the expected results are generally concluded to be reachable within the program's time frame.[30] Because more than 60 percent of the Army's advanced technology development efforts involve industry from their inception, established expectations provide valuable guidance and infrastructure for all involved parties.[31] In addition, Figure 3-4 depicts Army leadership's involvement with the S&T process to help establish priorities, to allocate funding based on those priorities, and thus to achieve program success.

Every year, the Army Training and Doctrine Command (TRADOC) and Head-

---

[26]Department of the Army. 2007. *Army Science and Technology Master Plan: Charting the Future of S&T for the Soldier.* Report by the Office of Deputy Assistant Secretary of the Army for Research and Technology. Washington, D.C.: Department of the Army. Available at http://www.carlisle.army.mil/dime/documents/JPLD_AY08_Lsn%207_Reading%204_ASTMP.pdf. Accessed September 2, 2010.

[27]Mary Miller. 2009. *Technology Transition—Lessons Learned.* Report produced by the Director for Technology, Office of Deputy Assistant Secretary for Research and Technology/Chief Scientist. Washington, D.C.: Department of the Army.

[28]Ibid.

[29]Department of the Army. 2008. *U.S. Army Weapons Systems 2009.* New York, N.Y.: Skyhorse Publishing.

[30]Department of the Army. 2007. *Army Science and Technology Master Plan: Charting the Future of S&T for the Soldier.* Report by the Office of Deputy Assistant Secretary of the Army for Research and Technology. Washington, D.C.: Department of the Army. Available at http://www.carlisle.army.mil/dime/documents/JPLD_AY08_Lsn%207_Reading%204_ASTMP.pdf. Accessed September 2, 2010.

[31]Mary Miller. 2009. *Technology Transition—Lessons Learned.* Report produced by the Director for Technology, Office of Deputy Assistant Secretary for Research and Technology/Chief Scientist. Washington, D.C.: Department of the Army.

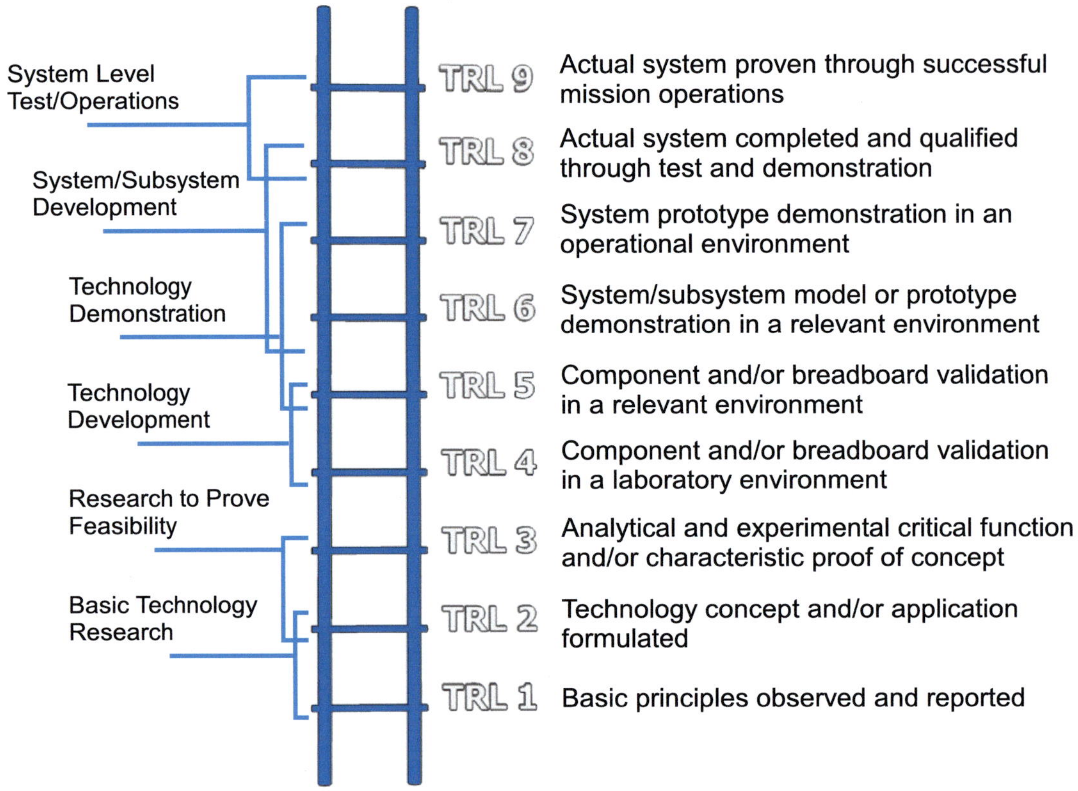

FIGURE 3-3
Department of Defense (DoD) hardware Technology Readiness Levels (TRLs). NOTE: Technology typically transitions out of science and technology and into an acquisition program after TRL 6. SOURCE: Based on information derived from the Department of Defense and the National Aeronautics and Space Administration.

quarters, Department of the Army, meet to review warfighter needs and establish priorities. That information is provided to a group of one-star general officers called the Warfighter Technical Council, which then submits its nominations for new ATOs, as well as revisions to existing ones. From there, nominations are reviewed by two-star general officers, all of whom have S&T oversight responsibilities. Finally, their recommendations are presented to a group of three-star and four-star General Officers called the Army Science and Technology Advisory Group, which ultimately signs off on or terminates the ATOs.[32] With this alignment between senior Army leadership, S&T leadership, and the front-line staff responsible for

[32]Mary Miller, Director for Technology, Office of Deputy Assistant Secretary for Research and Technology/Chief Scientist, Department of the Army.

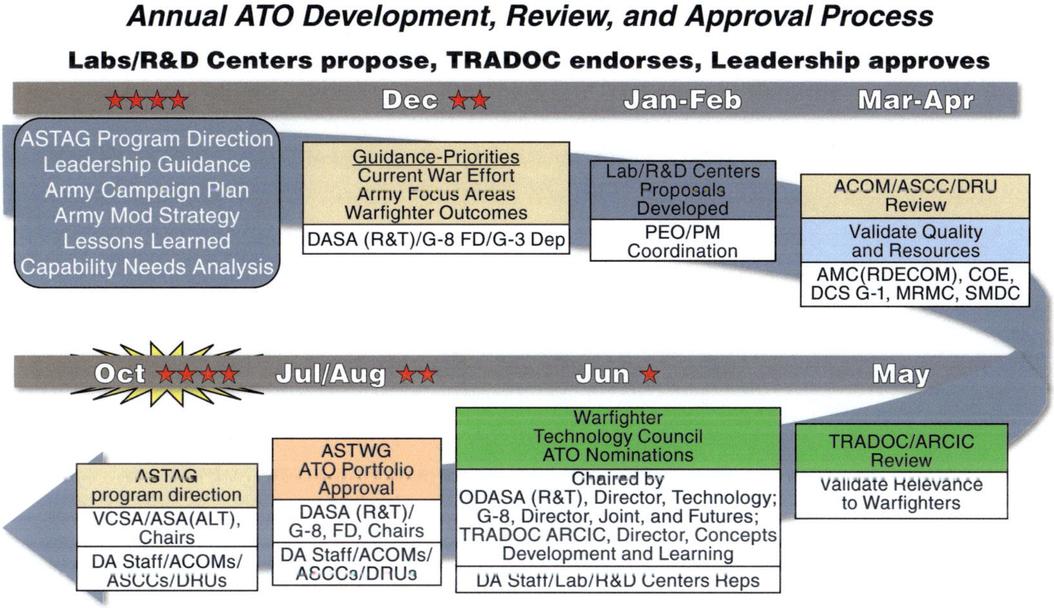

FIGURE 3-4
Army leadership involvement in the Army Technology Office (ATO) process. SOURCE: Mary Miller, Director for Technology, Office of Deputy Assistant Secretary, Department of the Army.

S&T development and transition, the ATO process has been successful, even when measured against the demands of a rapidly evolving battlespace.

FINDING 3-2
Successful technology transition is achieved by the participation of active senior service leadership, consistent priorities, and strong ties between commands responsible for science and technology, systems development and acquisition, and warfighting operations.

Dragon Eye

Responding to the U.S. Navy's call for more rapid fielding of technologies in the late 1990s, the head of the Naval Research Laboratory's (NRL's) Vehicle Research Section recommended seeking out simpler, more mature systems for rapid fielding. Specifically, his approach called for government laboratories to take the lead on prototyping and nailing down specifications, then engaging industry initially for build-to-print services, followed by spiral system upgrades to support continued capabilities growth.[33]

These principles were subsequently put into practice when the NRL was asked to provide an unmanned aircraft, dubbed Dragon Eye, with a smaller, less visible signature. Shown in Figure 3-5 is the hand-drawn sketch of the aircraft from an early meeting, outlining the required baseline configuration and proposed project plan.

Building on the fact that the NRL and similar organizations had been researching miniature unmanned air vehicle technologies for some time, Dragon Eye was scheduled in two phases: (1) a 12-month evaluation program including rigorous technical and operational evaluations, and (2) a 36-month development program to complete a comprehensive, mission-capable system design; to demonstrate aircraft prototypes; and to transition the aircraft to production by means of industry build-to-print services. Development was further partitioned to include an 18-month research and development phase including prototyping and the evaluation of industry best practices, followed by an 18-month period of refinement, system evaluation, and transition to production.[34]

Dragon Eye prototype evaluation began in 2001, and by 2004, production aircraft were deployed in both Iraq and Afghanistan. In its earliest configuration, the Dragon Eye aircraft offered man-portable, short-duration, "over-the-hill" surveillance capability. Incorporating proven, low-cost construction materials and commercially available sensors enabled the technology to be integrated, tested, and fielded expeditiously. This strategy also provided sufficient capabilities to address the user's basic requirements: for example, adequate visibility in low-light and daylight conditions.

Incremental upgrades allowed additional technologies to be more thoroughly vetted prior to deployment on the aircraft, reducing both risk and cost. For example, enhancements for increased endurance and sensors for day or night operation were not mature enough for deployment at the time of the urgent need for Dragon Eye. Instead, these capabilities were incrementally retrofitted once proven mature. To date, a total of more than 1,300 Dragon Eye aircraft, designated RQ-14A, have been deployed, confirming the merits of this technology development and acquisition approach.[35]

In the case of Dragon Eye, the customer ensured that system performance requirements were fully defined and understood by all parties at the outset, and

[33]Francis Klemm, Superintendent, Tactical Electronic Warfare Division, Naval Research Laboratory. 2010. "Dragon Eye—A Small UAV: From a Paper Sketch to an Operational System in 30 Months." Presentation to the committee, June 7, 2010.

[34]Ibid.

[35]Ibid.

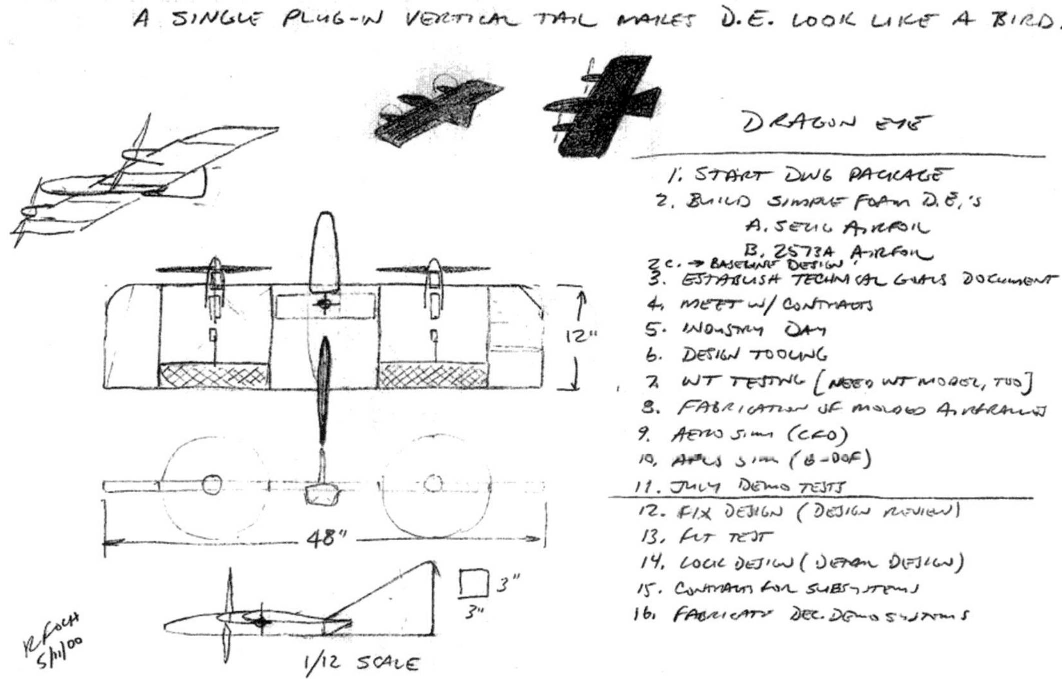

FIGURE 3-5
Original sketch of Dragon Eye. SOURCE: Francis Klemm, Superintendent, Tactical Electronic Warfare Division, Naval Research Laboratory. 2010. "Dragon Eye—A Small UAV: From a Paper Sketch to an Operational System in 30 Months." Presentation to the committee, June 7, 2010.

ongoing research was leveraged to apply proven mature technology solutions to those requirements. Likewise, spiral technology development was planned to ensure adequate time for system development, prototyping, and testing, as well as to allow evolutionary enhancements to the system post-production. Dragon Eye succeeded because a small, self-contained team of experts matched available mature technology with an urgent operational need to deliver an entire system quickly.

FINDING 3-3

A full understanding of the capabilities and limitations of the technology prior to committing to an acquisition program reduces the inclination to adopt unrealistic requirements.

EXAMPLES OF JOINT GOVERNMENT AND INDUSTRY COOPERATION

The Rapid Reaction Technology Office and the VADER System

Combining the best practices and core capabilities of multiple organizations can contribute to a faster and more effective process. For example, the Rapid Reaction Technology Office (RRTO), the organizational structure of which is shown in Figure 3-6, was established in 2006 under the Director, Defense Research and Engineering (DDR&E), within OSD.[36]

According to a recent report from the National Research Council (NRC), the small and agile RRTO has a specific charter to explore technologies that meet urgent needs of the warfighter and can be matured within 6 to 18 months for rapid transition into the field.[37] Table 3-1 details the RRTO's goals, by division.

After a given technology is combat proven, the RRTO places the program with the most appropriate government agency for maturing or application, as appropriate. Among the keys to the RRTO's success are experimentation, risk tolerance, and rigorous testing and prototyping.[38]

In 2009, the RRTO, in cooperation with the U.S. Department of Homeland Security, Customs and Border Protection, Northrop Grumman Corporation, and the Georgia Tech Research Institute, provided funding for a 5-night demonstration of the Vehicle and Dismount Exploitation Radar (VADER), a radar sensor.[39]

Integrated onto a Reaper unmanned aircraft owned by Customs and Border Protection, VADER was deployed along a 50-kilometer portion of the U.S. border with Mexico in Arizona.[40,41] VADER sensors collected data during the flight and transmitted them to several locations. The collaborative demonstration was a great success: VADER successfully identified suspicious activities during 4 of the 5 nights, and following up on the transmitted data, authorities were able to seize persons

---

[36]NRC. 2009. *Experimentation and Rapid Prototyping in Support of Counterterrorism.* Washington, D.C.: The National Academies Press.

[37]Ibid.

[38]Ibid.

[39]Benjamin Riley, Principal Deputy, Rapid Fielding Directorate, Office of the Under Secretary of Defense for Acquisition, Technology, and Logistics, Department of Defense. 2010. "Remarks to the National Research Council Air Force Preacquisition Technology Development Study Panel." Presentation to the committee, June 8, 2010.

[40]WTAM News. 2010. "Unmanned Drones to Patrol U.S.-Mexico Border." *WTAM 1100 News,* August 31. Available at http://www.wtam.com/cc-common/news/sections/newsarticle.html?feed=104668&article=7543084. Accessed September 3, 2010.

[41]Canwest News Service. 2009. "Predator Drones Patrolling Canada-U.S. Border." *National Post,* June 24. Available at http://www.nationalpost.com/news/canada/toronto/story.html?id=1727873. Accessed September 3, 2010.

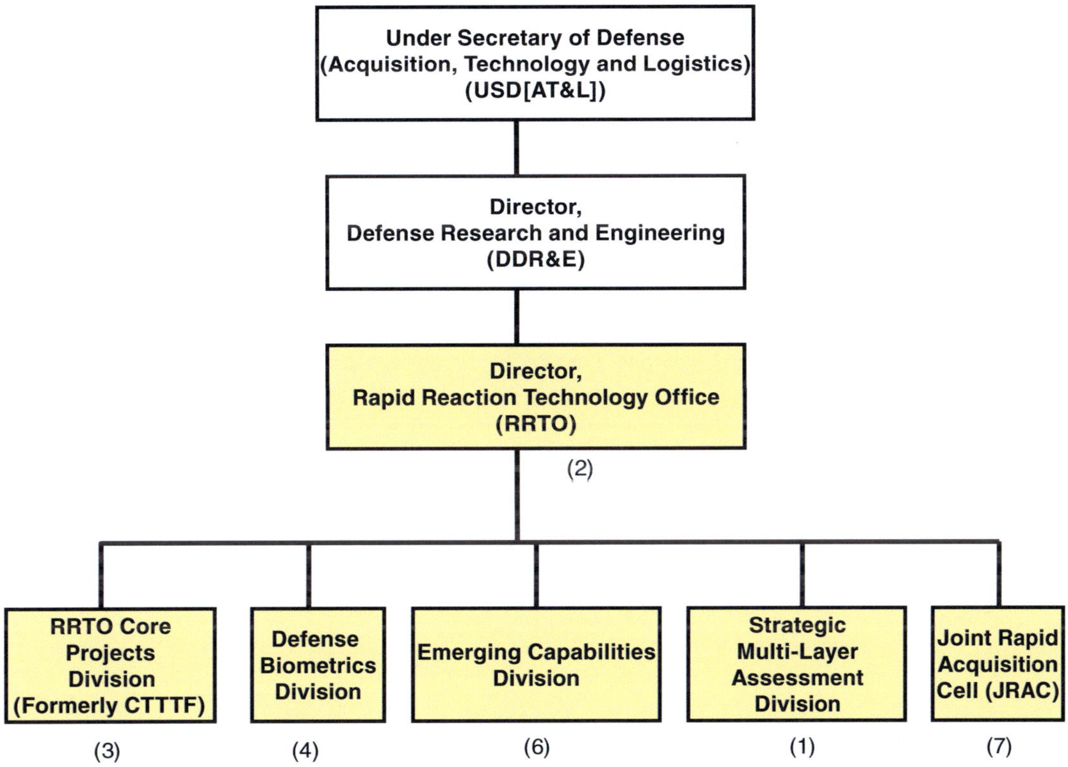

FIGURE 3-6
The organizational structure of the Rapid Reaction Technology Office (RRTO). SOURCE: Reprinted from Figure 2-1, National Research Council. 2009. *Experimentation and Rapid Prototyping in Support of Counterterrorism.* Washington, D.C.: The National Academies Press.

of interest as well as illegal narcotics.[42] In February 2010, Northrop Grumman announced that it had completed additional flight testing of the VADER system. According to the company:

> The VADER program is sponsored by the Joint Improvised Explosive Device Defeat Office (JIEDDO) and has been managed by both DARPA [Defense Advanced Research Projects Agency] and Defense Microelectronics Activity (DMEA) in conjunction with Army Intelligence and Information Warfare Directorate. DMEA awarded follow-on contracts to

[42]OSD. 2010. "RDT&E Project Justification: PB 2011." Available at http://www.dtic.mil/descriptivesum/Y2011/OSD/0605799D8Z_PB_2011.pdf. Accessed September 3, 2010.

TABLE 3-1 Goals and Focus of the Rapid Reaction Technology Office (RRTO), by Division

Project Division	Current Goals and Focus
Core Projects Division	Assumed responsibilities, functions, and projects from the RRTO's predecessor, Combating Terrorism Technology Task Force. Manages ongoing projects from the current areas of emphasis, which include multiple initiatives.
Defense Biometrics Division	Development of a defense-wide biometric capability that supports identity management, tactical biometrics and forensic applications, and force protection.
Emerging Capabilities Division	Supports the Joint Capabilities Integration Development System and acquisition processes. Develops prototypes with military utility in targeted areas of technologies and engages in activities for advanced capabilities, leveraging interagency cooperation and coordination.
Strategic Multi-Layer Assessment Division	Provides planning support to combatant commanders and coordinates with the Joint Staff and Strategic Command to support global mission analysis.
Joint Rapid Acquisition Cell (JRAC)	Addresses the rapid resolution of Joint Urgent Operational Needs Statements (JUONSs) and immediate warfighter needs. The JRAC monitors the status of validated JUONSs and assists in the resolution of issues that could result in mission failure or casualties.

SOURCE: Adapted from Table 2-1, National Research Council. 2009. *Experimentation and Rapid Prototyping in Support of Counterterrorism.* Washington, D.C.: The National Academies Press.

Northrop Grumman in January 2009 to improve the capability and support testing by the Army.[43]

In this case, the resources, expertise, and solutions of multiple organizations, from government, industry, and academia, merged to form a rapid and successful demonstration of a key developing technology. Moreover, the proposed solution applies not only to the homeland security application for which it was initially demonstrated, but to many other missions, domestically and internationally. As the Northrop Grumman announcement notes, "When deployed, VADER will provide U.S. Army ground commanders with real-time accurate Ground Moving Target Indicator data and Synthetic Aperture Radar imagery."[44]

DARPA's Adaptive Execution Office

The leadership of the Defense Advanced Research Projects Agency (DARPA) has created a new office, the Adaptive Execution Office (AEO), which is applying a best practice of "transition ability of DARPA programs." The AEO was modeled on both the Special Operations Command and Lockheed Martin Corporation's Advanced Development Programs, where fielding capability rapidly, for exceptional

[43]Northrop Grumman Corporation. 2010. "Northrop Grumman Successfully Demonstrates VADER Dismount Detection." Northrop Grumman News Release, February 16. Available at http://www.irconnect.com/noc/press/pages/news_releases.html?d=184378. Accessed August 9, 2010.

[44]Ibid.

needs, is the norm. The AEO director makes the point that "technology transition is not that hard but it takes passion." DARPA actively searches for bright, capable people to be program managers, and it expects them to commit to a 4-year tenure. Technology transition needs to be a program strategy "sooner rather than later" and is formalized in a DARPA Program Authorization Document (PAD) addendum. There is no one-size-fits-all technology transition process, no "cookie-cutter" approach—hence the word "adaptive" in the process. The AEO conducts critical self-assessments starting with the underlying science and using measurable metrics. To facilitate technology transition, DARPA then brings in program managers from each service to be bridges to DARPA programs. DARPA becomes an adviser once a program reaches Milestone B. The Air Force can benefit in terms of fielding and cost sharing by participating with DARPA and industry partners to meet operational needs.[45]

Ground Robotics Consortium

The Ground Robotics Consortium was created by the Joint Ground Robotics Enterprise (JGRE) in the OSD's office of Land Warfare and Munitions. The consortium provides a unique opportunity for nongovernmental organizations to participate in DoD research planning, producing a plan based on industry expert knowledge of evolving technology. It also allows the services to leverage companies' independent research and development funding through insights gained as a result of this mutual planning process, and it benefits the services by lowering barriers for small innovative companies to enter into the government acquisition process.

Since its creation in 1998, the Ground Robotics Consortium has attracted 205 members from for-profit and not-for-profit companies and academia. The JGRE provides oversight and guidance, conducts planning and budgeting, manages the acquisition process, coordinates with other organizations, and conducts source selection. The consortium provides liaison among all the members and with the JGRE, participates in the development of plans, and supports the JGRE subcommittees. Significant advantages to the military services include improved technology readiness, increased interoperability, and, most importantly, the accelerated transition of robotic innovations from commercial uses to warfighting applications.[46]

---

[45]Ellison Urban, Director, AEO, DARPA. 2010. Site visit by members of the committee, July 27, 2010, Washington, D.C.

[46]Ellen Purdy, Enterprise Director, Joint Ground Robotics, OUSD(ATL)/PSA/LW&M. 2008. "Ground Robotics. Industry/Academia Day. Government Technical Overview." Available at www.jointrobotics.com/GRE%20Concept%20Final%20ver3.ppt. Accessed September 3, 2010.

The National Small Arms Center

Established in 2004, the National Small Arms Center (NSAC) is managed by the Joint Services Small Arms Program Office at the Army's Armament Research, Development, and Engineering Center (ARDEC) at Picatinny Arsenal in New Jersey. With membership including government laboratories, engineering test sites, firing ranges, production facilities, industry, and academia, the NSAC's stated mission is "to mobilize the nation's intellectual and industrial resources to provide the best small arms systems to the nation's warfighters and law enforcement personnel."[47] As a hub to bring together these various stakeholders, the NSAC serves many roles. It is a center for research in the areas of small arms, providing consultation and training services to its members.[48]

Among these services are informational sessions, which enable government customers to outline their requirements before industry and academic members in order to set the thought process in motion before requests for proposals are issued. Likewise, industry and academia can meet to discuss customer priorities and explore how they might team to deliver the best ideas. In contrast, interested members can submit white papers on emerging technologies to assess interest by funding organizations in sponsoring additional research.

The NSAC utilizes the flexible Other Transactions Agreement (OTA) as its contracting vehicle for research and development, prototyping, and early production programs. OTAs, typically used for federal grants, are free of Federal Acquisition Regulation (FAR) clauses.[49] This enables the organization to draw on the expertise and innovation of nontraditional defense contractors such as small contractors and academic institutions that typically are unable to qualify for FAR-based contracts. In practice, significant collaboration and teaming occur between experienced DoD contractors and these nontraditional members, resulting in shared innovation and technology advancement.

For example, AAI Corporation is an industry member of the NSAC, with a long record of small arms research, development, and production. Its current NSAC programs, launched in 2010 and resulting from proposals attuned to the group's current interest in the areas of small arms lethality and fire control, include the following:

[47]National Small Arms Center. 2005. *Why Join National Small Arms Technology Consortium.* Available at http://www.nationalsmallarmscenter.com/public_docs/why_join_EME.ppt. Accessed September 3, 2010.

[48]Ibid.

[49]National Small Arms Center. 2010. *Public Documents Center. Other Transaction Agreements.* Available at http://nationalsmallarmscenter.com/public-documents/C33/. Accessed September 3, 2010.

- A combined lethal/nonlethal projectile, which would enable forces to transition seamlessly from a nonlethal projectile for crowd control to a lethal projectile for direct threats;
- An enhanced fragmentation warhead; and
- Improved fire control and power management.

Members of the NSAC benefit from continuous information sharing among stakeholders, an emphasis on long-term thinking about warfighter needs and the enabling technologies of the future, and the more flexible OTA contracting vehicle that encourages teaming and creativity.

FINDING 3-4
Collaborative practices between government agencies and industry can lead to successful technology insertion.

Sierra Nevada Corporation and the Commercial Space Sector

Industry succeeds best when it leverages the best skills and expertise available from all sources to feed technology development. In February 2010, Sierra Nevada Corporation's (SNC's) Space Systems Group was selected to receive a portion of the $50 million Commercial Crew Development stimulus funding provided by the U.S. government through NASA. The focus of this initiative is incentivizing the private sector to develop and demonstrate human spaceflight capabilities. Scheduled to complete Phase I in December 2010, SNC is responsible for the following:

- *Delivering a program implementation plan for a fast-tracked, streamlined contract vehicle based on payment for performance.* This approach is intended to reduce cost for all parties while allowing industry partners to contribute internal research and development dollars to the endeavor.
- *Completing a space vehicle review of its aeroshell tooling solution.* This review comprised 40 subsectors, including both mechanical and computer models. SNC passed this review in early 2010.
- *Demonstrating the space vehicle prime motor's functionality and capacity for reuse.* To reduce risk, the contract stipulates that SNC will receive additional funding only if the motor operates as intended.
- *Completing a primary starter test to demonstrate that systems will operate according to program requirements in the unique environment of space.*

SNC took several steps to ensure success in the space domain and, ultimately, success of this program.[50] It combined its own knowledge of government procurement with the specialized technical expertise of recent acquisitions to bring the best possible set of capabilities to the customer.[51] SNC is also leveraging proven, mature technologies to reduce overall program risk. Its product development model was accepted by NASA and has not only reduced risk but also allowed SNC to meet the first milestones within budget and on time. In this manner, NASA has gained trust in SNC's capabilities. This example underscores that technical skills, systems integration prowess, and knowledge of the acquisition process are all essential to delivering on customer requirements.[52]

The Naval Center for Space Technology

The NRL's Naval Center for Space Technology achieves a synergy with acquisition and industry that could serve as a model for other technology developers. The Naval Center for Space Technology has built a 45-year record of successful space systems development, with a number of impressive accomplishments. Some notable examples of its successful technology development and/or prototyping include the following:

- *Payload Data Management System:* A high-risk spacecraft avionics system developed in partnership with industry and subsequently transitioned.
- *TacSat-1 and -2:* TacSat-1, a signals intelligence satellite, developed and launched in less than a year, and for less than $15 million.
- *Oceanographic and meteorological payloads.*

Keys to the success of the Naval Center for Space Technology include the following:

- The empowering of small and accountable development teams,
- The augmenting of government laboratory expertise with industry capabilities, and
- The sharing of the scientific and technology development resources of a

[50]SNC. 2010. "SNC Receives Largest Award of NASA's CCDev Competitive Contract." Sierra Nevada Corporation Press Release. Available at http://www.sncorp.com/news/press/pr10/snc_ccdev_space-news.shtml. Accessed August 10, 2010.

[51]SNC. 2010. "Sierra Nevada Corporation Selected Under NASA's Human Space Transportation Program: SNC Dream Chaser™ Space Program to Provide Commercial Crew Capability." Sierra Nevada Corporation Press Release. Available at http://www.sncorp.com/news/press/pr10/snc_hstp.shtml. Accessed August 10, 2010.

[52]Mark Sirangelo, Corporate Vice President, Sierra Nevada Corporation, Space Systems Group. 2010. "SNC Space Systems." Presentation to the committee, July 6, 2010.

highly ranked laboratory—the Johns Hopkins University Applied Physics Laboratory (JHU/APL).

More recently, the Naval Center for Space Technology led the Integrated Government-Industry System Engineering Team (ISET) that developed TacSat-4. The spacecraft was built by the NRL and the JHU/APL, to standards developed collaboratively by the ISET. This collaborative approach represents a best practice for transitioning technology into operational application within tight constraints in terms of cost, schedule, and technical performance.

EXAMPLES OF INDUSTRY BEST PRACTICES

High Technological/Manufacturing/Integration Readiness Levels Pay Off: Ford, Jaguar, and Adaptive Cruise Control

The commercial automotive industry is akin to the defense industry in the sense that the former also must embrace the principles of speed to market and lean, affordable product development. Adaptive cruise control (ACC) was introduced in 1999 by the Ford Motor Company on the Jaguar automobile. ACC development began under the Program for European Traffic with Highest Efficiency and Unprecedented Safety (PROMETHEUS), which was intended to promote collaboration among vehicle manufacturers to garner improvements in traffic flow and safety.[53] As the U.S. Department of Transportation's Federal Highway Administration explains:

> PROMETHEUS was started in 1986 and was initiated as part of the EUREKA program, a pan-European initiative aimed at improving the competitive strength of Europe by stimulating development in such areas as information technology, telecommunications, robotics, and transport technology. The project is led by 18 European automobile companies, state authorities, and over 40 research institutions. The budget for the project is over $800 million and the project is scheduled to last seven years. PROMETHEUS is a pre-competitive research project, with the output being a common technological platform to be used by the participating companies once the product development phase begins.[54]

With this initial development already complete, Ford undertook its own internal technology development program in 1993. By 1995, the team had achieved TRL 5 maturity. Just a year later, ACC was launched formally after being demonstrated

[53]PR Newswire. 2010. "Jaguar Teams with Delphi to Introduce Adaptive Cruise Control." Available at http://www.prnewswire.co.uk/cgi/news/release?id=24470. Accessed August 9, 2010.

[54]U.S. Department of Transportation, Federal Highway Administration. 2010. "European ATIS Projects/Systems." Available at http://www.fhwa.dot.gov/tfhrc/safety/pubs/95153/sec5/body_sec5_01_04.html. Accessed August 10, 2010.

at TRL 8. In 1999, the company successfully introduced ACC to the marketplace on its Jaguar luxury vehicle.[55] In this example, the company harnessed a technology that already had been validated through extensive industry review and integrated with a larger system. The result was a successful product launch in a streamlined time frame.

FINDING 3-5
Decoupling technology maturation and system development has been proven to reduce overall risk dramatically.

There are instances in which the military services have sought to learn from industry in order to implement similar models. Today's technology transition leaders invoke a collaborative, open atmosphere of innovation akin to the ideas of open-source technology architecture or free-market economics—namely, that free interplay among stakeholders results in a better outcome for all parties. For example, at the University of California, Berkeley, Dr. Henry Chesbrough pioneered the concept of "open innovation," which encourages both the internal and external sharing of information and ideas as the true basis for success in innovation. The theory further contends that organizations which espouse "closed" innovation will miss crucial opportunities for success because ideas do not fall neatly into their established areas of operation.[56]

The Innovation Culture at 3M

The U.S. Army's Tank Automotive Research, Development, and Engineering Center (TARDEC) adopted an exchange program with commercial industry pioneer 3M to identify and harness best practices for technology development and innovation.[57]

A case study by the William F. Achtmeyer Center for Global Leadership at the Tuck School of Business at Dartmouth describes the 3M culture of "intrapreneurship." The study notes that nearly 35 percent of 3M's total sales for the year 2000 came from products introduced within the previous 4 years, showing the organi-

[55]Michael Sullivan, Director, Acquisition and Sourcing Management Team, U.S. Government Accountability Office (GAO). 2010. "Survey of GAO Studies and Findings." Presentation to the committee, April 21, 2010.

[56]For additional information on open innovation, see http://openinnovation.haas.berkeley.edu/openinnovation.html. Accessed October 27, 2010.

[57]Thomas Gehring, Program Manager, 3M Industrial and Transportation Business. 2010. "3M Innovation Story." Presentation to the committee, June 7, 2010.

zation's speed of innovation.[58] The company employs a rigorous gate process as a data-driven tool for the mid-program assessment of technology maturity and to drive investment decisions. Figure 3-7 shows a variation of this process developed for the TARDEC.[59]

Other drivers of innovation include the company's "15 percent option," which enables employees to spend up to 15 percent of their work hours pursuing self-driven development projects.[60] While the selection of these projects centers on the employee's interests and technical specialty, they often venture into new technology areas or markets.

Additionally, as with other cases discussed previously in this study, 3M strives to ensure that technology development and product development are never done concurrently. That is, only after a new technology is developed to a high Technology Readiness Level and Manufacturing Readiness Level (MRL) does 3M consider it suitable for incorporation into new product development. For 3M, the simultaneous development of new technology and new products is considered unacceptable.[61]

The U.S. Army has adopted these lessons from 3M. The TARDEC's Research Business Group independently benchmarked 3M's technology roadmapping, project portfolio management, and stage-gate project management processes. These best practices were used to bolster TARDEC's internal processes, and two pilot projects were selected for evaluation in 2010, with full implementation planned for 2011.[62]

The importance of evaluating the TRL, MRL, and Integration Readiness Level together as a measure of overall technology maturity was highlighted previously in this report. Indeed, recent changes in DoD Instruction 5000.02 substantially increase the emphasis on technology development and maturation. The pre-systems acquisition phase between Milestones A and B is now focused on reducing technology risk prior to contracting for Engineering and Manufacturing Development. This new direction, as set forth in the Weapon Systems Acquisition Reform Act of 2009 (Public Law 111-23) and in DoD Instruction 5000.02, is fully consistent with

---

[58]William F. Achtmeyer Center for Global Leadership at the Tuck School of Business at Dartmouth. 2002. *3M Corporation.* Available at http://mba.tuck.dartmouth.edu/cgl/downloads/20002_3M.pdf. Accessed August 10, 2010.

[59]Thomas Gehring, Program Manager, 3M Industrial and Transportation Business. 2010. "3M Innovation Story." Presentation to the committee, June 7, 2010.

[60]William F. Achtmeyer Center for Global Leadership at the Tuck School of Business at Dartmouth. 2002. *3M Corporation.* Available at http://mba.tuck.dartmouth.edu/cgl/downloads/20002_3M.pdf. Accessed August 10, 2010.

[61]Thomas Gehring, Program Manager, 3M Industrial and Transportation Business. 2010. "3M Innovation Story." Presentation to the committee, June 7, 2010.

[62]Ibid.

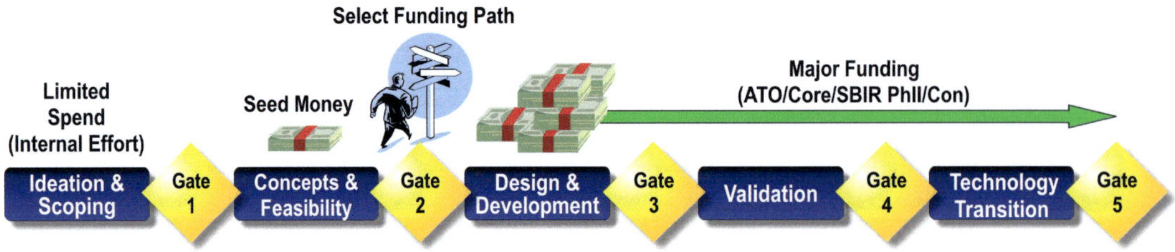

FIGURE 3-7
The stage-gate process of the U.S. Army's Tank Automotive Research, Development, and Engineering Center (TARDEC). SOURCE: Thomas Gehring, Program Manager, 3M Industrial and Transportation Business; and Heather Molitoris, U.S. Army-TARDEC. 2010. "Technology Development and Innovation at 3M Company." Presentation to the committee, June 7, 2010.

the best practices of industry innovation leaders like Ford/Jaguar and 3M. Their experience with the astute application of TRL assessment methods shows clearly that the TRL is of critical importance. Assessing readiness level in an objective and analytically rigorous manner is a start. Equally important, though, is maintaining the organizational discipline necessary to follow through on those assessments by not proceeding through decision gates until the required maturity is reached.

FINDING 3-6
Independent, rigorous, analytically based characterization of Technology, Manufacturing, and Integration Readiness Levels will lead to higher confidence and a greater likelihood of successful outcomes.

Technology Networks at Raytheon

Raytheon Company employs a system of technology networks to facilitate communication and collaboration across various locations, businesses, and engineering functional organizations. The technology networks are mandated and supported at the highest levels of the company. In fact, each network is led by one of the company's senior technologists serving as the chairperson, supported by a council of technologists and a facilitator.

Five technology networks are built around Raytheon's core technical competencies; these networks are further divided into numerous technology interest groups (TIGs), which can be formed at the initiative of employees, or chartered by the company to address strategic business objectives. Each TIG is supported by a core set of information-sharing tools that are provisioned at the corporate level.

Each technology network hosts an annual symposium offering several technical tracks and tutorials for the 200 to 500 attendees. These symposia bring together colleagues from multiple locations to discuss common interests and issues, share solutions, and build relationships that facilitate future cooperation. Customers and suppliers are often invited to speak to and to interact with the participants. In addition, 1- to 3-day workshops are conducted with fewer attendees and a more focused technical scope, addressing a topic of strategic or competitive interest to a program, a product line, or the company as a whole. Workshops culminate with the development of recommendations for Raytheon's approach to the selected topic. Each TIG hosts informative seminars to cover a spectrum of special-interest topics from new technologies and processes to new tools and supplier products, either available internally or elsewhere in industry.

Raytheon's robust technology networking infrastructure provides a conduit for temporary alignment among people with diverse interests, roles, and responsibilities. When they return to their own locales and work areas, that information is redistributed even further among the workforce—speeding communication, increasing technology reuse, and ultimately delivering greater value to the customer. In addition, this example illustrates that technology can be conceptualized and advanced only to a certain point within one segment of an organization. Once it is opened to the larger organization for debate, evaluation, and validation through peer reviews or gate reviews, the technology is subject to even greater and more rapid opportunities for advancement, as well as successful implementation into a larger variety of applications.

A single example speaks to the power of this kind of networking in a large organization and the direct benefit that accrues to both the business and the customer. A particular product line was about to deliver a command-and-control facility to a customer; in the process of checking to see that all of the requirements had been met, it was discovered that a particular capability had not been certified, and the program chief engineer was uncertain how it should be done. An urgent e-mail was sent out to the appropriate technology network asking for domain experts in this particular topic area. Within 24 hours, six experts had been identified across the company, one of whom was located at the site in need of the help. The appropriate testing and certification were quickly completed, and the facility was delivered on time and in full compliance to a happy customer.

This communication and dissemination example, in which technology developers collaborate effectively to advance the organization's objectives, would have applicability and value to the science and technology efforts of the Air Force.

CONCLUDING THOUGHTS

Several common threads run through the success stories outlined in this chapter. They include the following: the willingness to accept a good "80 percent solution" in months rather than a perfect answer years in the future; an objective and accurate assessment of technology maturity; Integrated Product Teams that bring together the right people with best-of-breed technologies and processes; accountability for the forward trajectory of programs; and the dedication of adequate funding. In addition, the principles represented by the "Three Rs"—Requirements, Resources, and the Right People—are applicable to each best practice. Regardless of whether the procurement is evolutionary or revolutionary and regardless of whether it originates within the commercial or the DoD realm, all "Three Rs" continue to emerge as being critical to preacquisition technology development and maturity as well as to ultimate program success.

4

The Recommended Path Forward

The preceding chapters provide the analytical foundation for recommendations contained in this chapter. Since the inception of the United States Air Force, its technological edge has been crucial to its success. This edge has been endangered over the past two decades, for the reasons cited in Chapter 2. Chapter 3 discusses characteristics common to organizations that do an exceptional job of specifying, developing, testing, and inserting new technology into their products or systems. These common characteristics, aligned with the "Three Rs"—(1) Requirements, (2) Resources, and (3) the Right People—the framework developed by the committee (see Box 1-2 in Chapter 1), can allow the Air Force to improve its ability to specify, develop, test, and insert new technology into its systems. This framework calls for clear, realistic, stable, trade-off-tolerant, and universally understood *requirements*; the *resources* needed to accomplish the job (including funding, together with policy and processes tailored for rapid technology insertion); and the *right people* (both in the government and as contractors) in the workforce and in charge.

Comparing the best practices exemplified in Chapter 3 with the shortfalls discussed in Chapter 2 leads to the seven key issues that the committee believes the Air Force must address in order to leverage quickly, correctly, and affordably the advanced technologies necessary to maintaining its warfighting edge. Table 4-1 summarizes these seven key issues, categorizes them in terms of the "Three Rs," and for each issue identifies which specific criteria in this study's statement of task (see Box 1-1) are addressed by one or more of the recommendations of the committee. The recommendations themselves are presented below in this chapter.

The seven key issues summarized in Table 4-1 are described in more detail

TABLE 4-1 Committee Recommendations Associated with the Seven Key Issues Identified in This Report

Key Issue	Statement of Task Criteria					
	Workforce	Organization	Policies	Processes	Resources	Legislatio
Requirements						
1. Freezing requirements too early or too late in the technology development phase can lead to a mismatch between technology-enabled capabilities and requirement expectations that significantly reduces the probability of successful technology transitions.			X	X		
Resources						
2. The lack of an Air Force-level science and technology strategy leads to AFRL efforts that may not support desired strategic Air Force capabilities, and to the fragmented prioritization and allocation of 6.4 technology transition funds.		X	X	X	X	
3. Current Air Force funding and business practices for Pre-Milestone B activities are inconsistent with DoD Instruction 5000.02			X	X	X	
4. Technology Readiness Levels must be accurately assessed to prevent programs from entering the Engineering and Manufacturing Development phase with immature technology.	X		X	X	X	X
5. Developing technologies and weapon systems in parallel almost inevitably causes cost overruns, schedule slippage, and/or the eventual reduction in planned capabilities.			X	X	X	
6. Weak ties and lack of collaboration within and between government and industry lead to lack of awareness of government priorities and of industry's technology breakthroughs.	X	X	X	X	X	
The Right People						
7. A much reduced and inexperienced Development Planning workforce has weakened the technology transition bridge between laboratories, Product Centers, and Major Commands.	X	X			X	

NOTE: The X's in the table indicate that one or more of the recommendations provided in this chapter address a pa ticular criterion. The recommendations address all criteria in the statement of task.

below. The description of each issue is followed by relevant findings (numbered in parentheses to match their numbering as presented in Chapter 2 or 3), and a recommendation associated with that issue. The chapter contains 17 of the study's findings and the 7 recommendations of the study.[1]

KEY ISSUE 1

Freezing Requirements Too Early or Too Late in the Technology Development Phase Can Lead to a Mismatch Between Technology-Enabled Capabilities and Requirement Expectations That Significantly Reduces the Probability of Successful Technology Transitions

Imposing a large and rigid set of requirements at the outset of the technology development phase can create false expectations among stakeholders, who may assume that technology "miracles" will occur, enabling the desired capabilities. In such cases, rather than reconsidering requirement expectations when technologies do not live up to early promises, stakeholders holding to an inflexible "I want what-I-want" position force programs to take on significant cost, schedule, and performance risks in pursuit of technologies that may never mature. Conversely, programs that freeze requirement too late in the technology development phase—for example, after System Requirements Review—fail to provide stable, objective goals for assessing technology maturity and for containing cost and schedule slippage. Successful programs "viciously manage" requirements, beginning technology development with a reasonable and flexible set of commonly understood requirements.[2] In these success stories, acquisition executives such as Product Center Commanders and Program Executive Officers ensure that a program's cost-capability information is correct and current. As the true life-cycle costs and capabilities of new technologies become known, Major Command (MAJCOM) customers are willing to trade off requirement desires against the cost, benefits, and readiness of new technologies in order to achieve an optimum set of capabilities in a reasonable time and at an affordable cost.

[1]The recommendations in this report apply to each of the three operational domains of the Air Force: air, space, and cyberspace. At the same time, each domain is unique due to its particular characteristics and the unique environments in which it operates. Several other findings besides those given here appear separately in Chapters 2 and 3.

[2]Douglas Shane, President, Scaled Composites. 2010. "Rapid Prototyping at Scaled Composites." Presentation to the committee, May 13, 2010.

FINDING (2-5)
After System Requirements Review, stable requirements and a well-defined operational environment are essential to successful technology insertion.

RECOMMENDATION 4-1
To ensure that technologies and operational requirements are well matched, the Air Force should create an environment that allows stakeholders—warfighters, laboratories, acquisition centers, and industry—to trade off technologies with operational requirements prior to Milestone B.

KEY ISSUE 2

The Lack of an Air Force-Level Science and Technology Strategy Leads to AFRL Efforts That May Not Support Desired Strategic Air Force Capabilities, and to the Fragmented Prioritization and Allocation of 6.4 Technology Transition Funds

If Air Force and industry efforts are to be focused on critical technology needs, then a process must exist at the corporate Air Force level to prepare and promulgate an Air Force science and technology (S&T) strategy. Unlike the Navy, whose Future Naval Capabilities process yields a strategic Navy-wide S&T plan overseen by a corporate structure consisting of research and development (R&D), program management, and operational stakeholders, the current Air Force process allows individual stakeholders, such as the Air Force Research Laboratory (AFRL) and the MAJCOMs, to develop and fund their own priorities. Historically, the Air Force relied on mechanisms like Applied Technology Councils (ATCs) to bring together technology developers (i.e., AFRL), the operational community (i.e., MAJCOMs), and the acquisition community (e.g., the Product Centers) to reach agreement on which technology developments were most needed and would therefore be funded and incorporated into programs of record. Unfortunately, the Air Force has in some cases allowed ATCs to atrophy, weakening a viable process responsible for making strategic technology transition funding decisions. In addition, the Air Force Product Center Development Planning Organizations (XRs) have a 6.4 Program Element (PE) for Requirements Analysis and Maturation (RAM). But unlike research, development, test, and evaluation (RDT&E) funding, which is prioritized and allocated by a single PE panel, 6.4 funding is managed by a diverse set of panels. Although this approach increases the likelihood that individual MAJCOM needs are met, it does not necessarily result in a global set of technology transition investments that address strategic Air Force priorities. In addition, once technology transition funds are distributed to the MAJCOMs, they tend to use the funds to solve near-term problems that may be inconsistent with strategic Air Force priorities.

FINDING (2-8)
The Air Force lacks an effective process for determining which technology transitions to fund.

FINDING (2-9)
Although the Air Force Chief Scientist has developed an "art of the possible" science and technology strategic plan for the 2010 to 2030 time frame, there exists no Air Force-level unifying strategy, inextricably linked to operational requirements, to guide decision making for science and technology investments.

FINDING (2-10)
Successful technology development and technology transition require (1) integration of warfighter requirements with science and technology investments and systems acquisition strategies, and (2) close collaboration among all government and industry partners.

FINDING (2-11)
MAJCOM ownership of Budget Category 4 Program Elements and the current Air Force Budget formulation process do not provide development planners with sufficient priorities for execution of maturation funding. At a higher level, the Air Force lacks an overarching strategy for technology development, or a process that involves key decision makers. As a result, there is no integrated view of warfighter needs and technological possibilities, and there is inadequate guidance for determining what technology transitions to fund.[3]

RECOMMENDATION 4-2
To enable (1) a more disciplined decision-making process and (2) a forum in

[3]Budget Activity 4, Advanced Component Development and Prototypes (ACD&P): Efforts necessary to evaluate integrated technologies, representative modes, or prototype systems in a high-fidelity and realistic operating environment are funded in this budget activity. The ACD&P phase includes system-specific efforts that help expedite technology transition from the laboratory to operational use. Emphasis is on proving component and subsystem maturity prior to integration in major and complex systems, and may involve risk-reduction initiatives. Program elements in this category involve efforts prior to Milestone B and are referred to as advanced component development activities and include technology demonstration. Completion of Technology Readiness Levels 6 and 7 should be achieved for major programs. Program control is exercised at the program and project level. A logical progression of program phases and development and/or production funding must be evident in the FYDP. *DoD Financial Regulation*, Volume 2B, Chapter 5, June 2004.

which all stakeholders—those from the science and technology (S&T), acquisition, and warfighting MAJCOM communities—can focus their attention jointly on critical technology development questions and then make tough strategy and resource calls efficiently at a level where the decisions are most likely to stick, the Air Force should consider adopting a structure similar to the Navy's S&T Corporate Board and Technology Oversight Group and the Army Technology Objectives Process and Army S&T Advisory Group. A committee-developed notional organization for Air Force consideration (Figure 4-1) addresses this potential and is tailored to Air Force missions and organization. In addition, the Air Force should consider allocating funding for technology development, including funding for 6.4, or advanced component development and prototypes, to the Air Force Materiel Command and Air Force Space Command, unless precluded by law from doing so.

In the opinion of the committee, this recommendation to add another organization to the Headquarters Air Force does not diminish the statutory and mission responsibilities of the Assistant Secretary of the Air Force (Acquisition) (SAF/AQ)[4] and is justified by the seriousness of the need. In the committee's judgment, no other approach would meet the need to bring together the S&T, acquisition, and warfighting MAJCOM communities at a level that could make the difficult decisions. The fundamental premise of Recommendation 4-2 is the importance of technology to the Air Force, as described in the introductory paragraphs of Chapter 1 and reiterated in the introductory statements in Chapter 2. Findings 2-8 and 2-9 identify significant shortfalls in decision making for Air Force technology development and transition—that is, the lack of a process for technology transition and, at a higher level, the lack of a service-wide unifying S&T strategy to guide investments—which, in the judgment of the committee, need to be addressed. The structure proposed in Recommendation 4-2 would give SAF/AQ greater leverage to ensure that the right technology is being developed, matured, and transitioned. Furthermore, the cross-domain character of technology development, addressed in Chapters 1 through 3 of this report, presents challenges that the recommended S&T Board could address efficiently with a diverse set of stakeholders at the table. Finally, given the ever-increasing complexity and budget implications of new weapons systems, in the opinion of the committee the status quo is not acceptable.

[4]The SAF/AQ's responsibilities are specified under the Goldwater-Nichols Department of Defense Reorganization Act of 1986 (Public Law 99-433) and Headquarters Air Force Mission Directive 1-10, April 8, 2009.

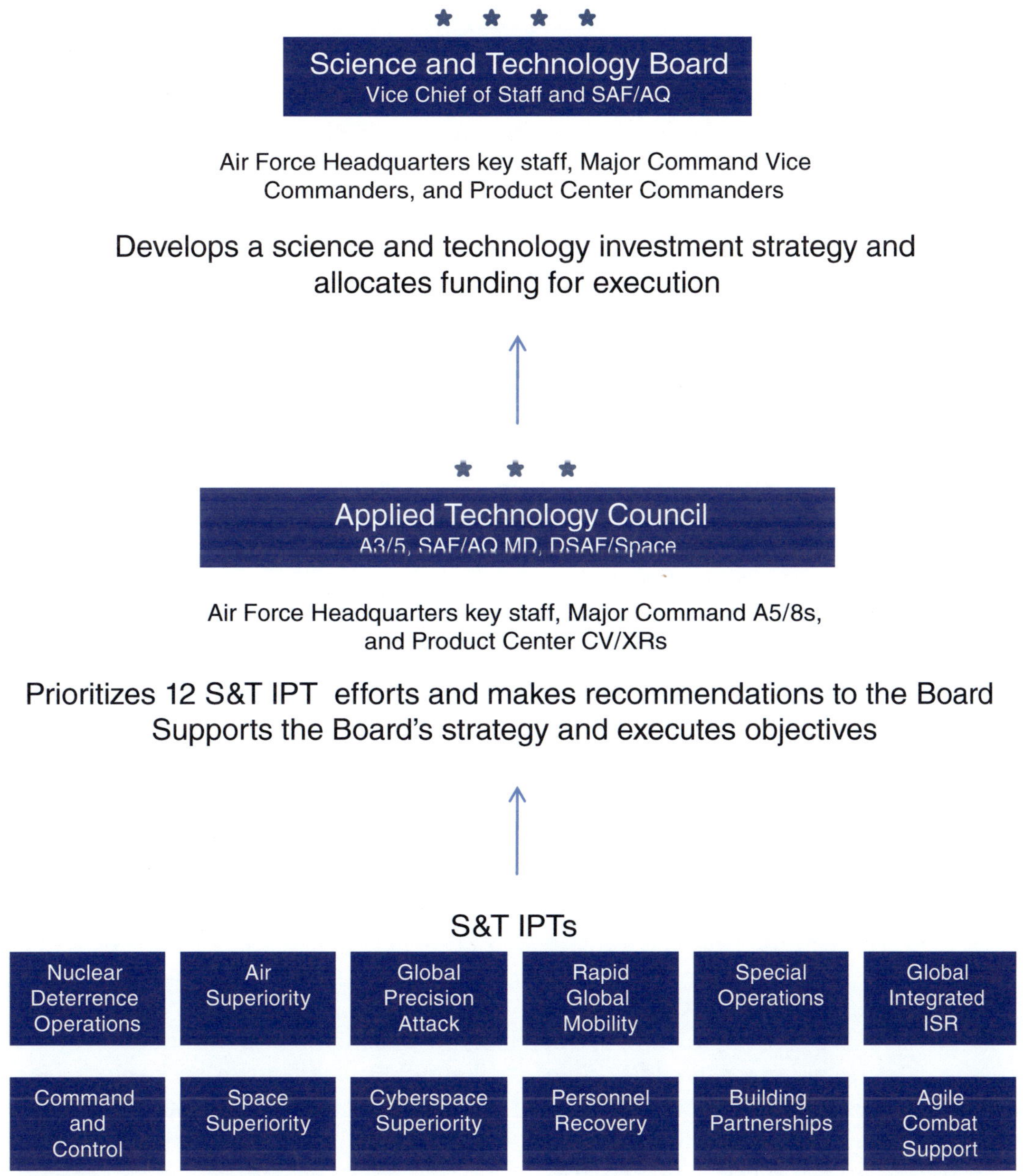

FIGURE 4-1
Notional science and technology governance.

KEY ISSUE 3

Current Air Force Funding and Business Practices for Pre-Milestone B Activities Are Inconsistent with Department of Defense Instruction 5000.02

Department of Defense (DoD) Instruction 5000.02 specifically states that processes, reviews, and milestones should be tailored for different program circumstances.[5] However, the committee learned from numerous presenters that the acquisition community often treats DoD Instruction 5000.02 pre-Milestone B guidance as rigid, leading to long and sometimes costly technology insertion campaigns. For example, current policy requires Preliminary Design Reviews prior to Milestone B, even though in some cases (e.g., competitive pre-Milestone B contracts) the Engineering and Manufacturing Development (EMD) contractor has not been selected, and detailed system design information does not exist. In addition, expensive and lengthy competitive prototyping efforts are sometimes implemented to comply with acquisition directives when the best prototype may be known in advance of the competitive prototyping procurement.

FINDING (2-1)
The Air Force competitive prototyping policy, AFI 63-101, lacks a waiver process for competitive prototyping.

FINDING (3-1)
Tailored processes can enable rapid technology insertion.

FINDING (3-3)
A full understanding of the capabilities and limitations of the technology prior to committing to an acquisition program reduces the inclination to adopt unrealistic requirements.

RECOMMENDATION 4-3
Since DoD Instruction 5000.02 incorporates increased pre-Milestone B work, the Air Force should bring pre-Milestone B work content back into balance with available resources by some combination of (1) DoD Instruction 5000.02 tailoring and/or (2) additional expertise, schedule, and financial resources. Examples of expanded content include competitive prototyping, demonstrating technology in operationally relevant environments, and completing preliminary design prior to Milestone B.

[5]DoD. Department of Defense Instruction 5000.02. December 8. Available at http://www.dtic.mil/whs/directives/corres/pdf/500002p.pdf. Accessed January 29, 2011.

KEY ISSUE 4

Technology Readiness Levels Must Be Accurately Assessed to Prevent Programs from Entering the Engineering and Manufacturing Development Phase with Immature Technology

The existence and definition of Technology Readiness Levels (TRLs) are now commonly understood across the acquisition community. However, methods for assessing TRL maturity are not as well understood. This is a particular problem for the evaluation of technologies at the transition "tipping point," where TRL guidance requires agreement on the definition of an "operationally relevant" environment. In many cases there can be considerable disagreement among well-intentioned experts on "relevance" criteria. Congress, through the Weapon Systems Acquisition Reform Act (WSARA) of 2009 (Public Law 111-23) assigned the Director, Defense Research and Engineering (DDR&E), the responsibility of conducting independent TRL assessments for selected Major Defense Acquisition Programs. In addition to concerns over the availability of DDR&E resources to accomplish this tasking, the ability to conduct independent TRL assessments needs to be vested in the Air Force acquisition system, initially to support non-Major Defense Acquisition Programs, and eventually to support all programs with DDR&E and congressional approval.

FINDING (3-6)
Independent, rigorous, and analytically based characterization of Technology, Manufacturing, and Integration Readiness Levels will lead to higher confidence and a greater likelihood of successful outcomes.

RECOMMENDATION 4-4
Knowledgeable, experienced, and independent technical acquisition professionals outside the program office should conduct technology, manufacturing, and integration assessments using consistent, rigorous, and analytically based standards. While WSARA requires this effort to be executed at the Office of the Secretary of Defense (OSD) level, this organic capability needs to be developed and assigned to the Air Force Materiel Command (AFMC) and the Air Force Space Command (AFSPC). Once this capability has been effectively demonstrated by the Air Force, legislative relief should be sought.

KEY ISSUE 5

Developing Technologies and Weapon Systems in Parallel Almost Inevitably Causes Cost Overruns, Schedule Slippage, and/or the Eventual Reduction in Planned Capabilities

The committee heard examples from numerous government and industry presenters about the pitfalls of entering the weapons system development phase while continuing to mature underdeveloped technologies. Because technology development often requires sudden moments of inventive inspiration, it is impossible to predict whether a technology will mature in time to meet important programmatic milestones. And even if the inspired moment does occur in time, the technology may not ultimately provide the desired performance or increased capability. It is therefore essential that unproven technologies be given sufficient time and resources to demonstrate their potential *before* entering into an acquisition effort that relies on that technology to achieve cost or performance goals. If the technology fails to meet expectations in time, the program is on the road to cost overruns and schedule slippage, with the program office needing to work with the operational community to adjust (i.e., reduce) capability expectations, or to seek additional funds and time to mature the technology (e.g., see the subsection "A Case Study on the Importance of Ensuring Technological Readiness: The Joint Strike Fighter" in Chapter 2).

FINDING (2-4)
The absence of independent, rigorous, analytically-based assessments of Technology, Manufacturing, and Integration Readiness Levels will reduce the likelihood of successful program outcomes. Furthermore, despite the existence of clear and compelling examples to the contrary, the Air Force continues to initiate system acquisition prior to completing the required technology development.

FINDING (3-5)
Decoupling technology maturation and system development has been proven to reduce overall risk dramatically.

RECOMMENDATION 4-5
To increase the likelihood of acquisition success, the Air Force should enter Engineering and Manufacturing Development (Milestone B) only with mature technologies—that is, with technologies at TRL 6 or greater.

KEY ISSUE 6

Weak Ties and Lack of Collaboration Within and Between Government and Industry Lead to Lack of Awareness of Government Priorities and of Industry's Technology Breakthroughs

Laboratory, acquisition, and operational organizations in many cases pursue their own technology development and transition agendas. And nearly all organizations operate with only a modest understanding of industry investments in independent research and development (IR&D). Certainly "technology-push" efforts are always needed to capitalize on or respond to surprise breakthrough technologies, but the government must strive to strike a balance between "blue-sky" research and the "technology-pull" efforts driven by stated capability needs. The industrially funded Naval Research Laboratory (NRL) offers a useful example of tying laboratory technology-pull activities more closely to the acquisition and operational communities by requiring NRL researchers to compete for government funding with their industry counterparts. Collaborative government-industry forums, such as those sponsored by the Space and Missile Systems Center Development Planning organization, the Ground Robotics Consortium, and the National Small Arms Center, also serve to educate government decision makers about industry R&D plans and to motivate industry researchers to focus on technologies required to achieve desired future capabilities. Workforce training and incentives may be needed to establish such forums.

FINDING (2-6)
Some important technology insertion efforts have failed to mature due to the lack of (or subsequent loss of) a specific targeted program of record—for example, a new engine technology being developed for a proposed aircraft. Thus, a successful and useful technology may go dormant until a new program can be identified to host it. In this manner, even valuable technology advancements that cannot be inserted in a timely way into a program of record might be relegated to the "Valley of Death."

FINDING (2-7)
The array of technology possibilities always exceeds the resources available to pursue them. One result is that the technology planning process tends to overcommit available resources and does not always ensure that every technology investment has an executable plan (with a corresponding budget) that enables near-term production readiness.

FINDING (3-2)
Successful technology transition is achieved by the participation of active senior service leadership, consistent priorities, and strong ties between commands responsible for science and technology, systems development and acquisition, and warfighting operations.

FINDING (3-4)
Collaborative practices between government agencies and industry can lead to successful technology insertion.

RECOMMENDATION 4-6
The Air Force should drive greater collaboration between warfighters (to include joint and coalition partners), laboratories, developers, and industry. One approach is to establish collaboration forums similar to the Ground Robotics Consortium and the Army Armament Research, Development, and Engineering Center's National Small Arms Center.

KEY ISSUE 7

A Much Reduced and Inexperienced Development Planning Workforce Has Weakened the Technology Transition Bridge Between Laboratories, Product Centers, and Major Commands

Historically, much of the Air Force responsibility for technology development and maturation rested with the Product Center Development Planning Organizations. Processes such as Vanguard linked laboratory, Product Center, and operational stakeholders to manage Air Force technology investments collaboratively. In the past two decades, Development Planning (DP) budgets were significantly reduced and eventually eliminated, leaving no organizations explicitly accountable for technology transition, no concentration of funds to mature and transition technology, and no repository to capture and pass lessons learned on to following generations. In addition, as the experience base dwindled, failures began to rise, spawning a vicious cycle of failure, distrust, and the addition of layer upon layer of expensive DoD and congressional oversight. Fortunately, recent investments in the restoration of Development Planning, including the release of new SAF/AQ policies, reconstitution of the XRs, and restoration of limited technology transition funding by Congress, are continuing. While the Air Force has begun to take steps to repair the damage done to the DP function, more needs to be done, including increasing funding for, and managerial emphasis on, the DP organizations and processes. Further, review of what was effective when the Air Force had a strong DP function would likely hasten its return.

FINDING (2-2)
Lack of trust and increasing oversight of Air Force technology development and acquisition by the Congress, OSD, and Air Staff are making successful program execution ever more difficult.

FINDING (2-3)
The decline of Development Planning and, in some quarters, the deterioration in the effectiveness of ATCs have greatly reduced the ability to integrate successfully the interests of warfighters, the S&T community, and acquisition leadership.

RECOMMENDATION 4-7
The Air Force should accelerate the re-establishment of the Development Planning organizations and workforce and should endow them with sufficient funds, expertise, and authority to restore trust in their ability to lead and manage the technology transition mission successfully.

CONCLUSION

From its inception, the Air Force has depended on advanced technology for an edge to overcome quantitative shortfalls—a comparative advantage that will likely become ever more important in a world of constrained defense budgets and diversified worldwide threats. Over the past two decades, the ability to specify, develop, test, and insert new technology into major Air Force systems was allowed to atrophy. More recently, the Air Force has recognized this deficiency and has started to reconstitute that capability; however, even more needs to be done. The recommendations in this chapter are intended to support and enhance that reconstitution effort and to help restore the Air Force's qualitative technical edge. It is crucial to recognize, however, that restoring that technological edge will require a reversal of the lost trust discussed earlier in this report. By breaking that cycle of mistrust and returning to the fundamentals of the "Three Rs"—Requirements, Resources, and the Right People—the Air Force can return to the days when its superb technological leadership set the example for others to follow.

Appendixes

Appendix A

Biographical Sketches of Committee Members

Richard V. Reynolds, *Chair*, Lieutenant General, United States Air Force (USAF; retired), is owner and principal of The VanFleet Group, LLC, an aerospace consulting company. He also serves as an independent director for Apogee Enterprises, Inc.; Barco Federal Systems, LLC; Allison Transmission, Inc.; the GE Rolls-Royce Fighter Engine Team, LLC; and EWA-GSI. Additionally he is an adviser to the USAF Heritage Program board of directors, president of the Air Force Museum Foundation, a trustee of the United States Air and Trade Show and Flight Test Historical Foundation, and secretary of Air Camp, Inc., and he serves on a number of other boards and committees in the local Dayton, Ohio, region. Prior to his retirement in 2005, General Reynolds was vice commander, Air Force Materiel Command, responsible for technological superiority, acquisition support, and testing, and sustainment of Air Force ground and airborne systems. He has commanded the Aeronautical Systems Center at Wright-Patterson Air Force Base, Ohio, and the Air Force Flight Test Center at Edwards Air Force Base, California. He has also served as Program Executive Officer for Airlift and Trainers in the Pentagon. General Reynolds is a graduate of the U.S. Air Force Test Pilot School, Class 79B, and has more than 25 years of experience in the research, development, testing, and evaluation of aeronautical systems. He was program director for several major weapon system acquisitions, including the B-2 Spirit. His logbook shows more than 4,000 flying hours in 67 different military and civil aircraft. Graduating in 1971 from the U.S. Air Force Academy with a Bachelor of Science degree in aeronautical engineering, General Reynolds holds a Master of Science degree in mechanical engineering from California State University and a Master of Arts degree in national security and strategic studies from the Naval War College. He is a Fellow of the Society of Experimental Test Pilots.

Donald C. Fraser, *Vice Chair*, NAE, has broad research management experience and is the founder and retired director of the Boston University Photonics Center. Dr. Fraser has had a distinguished career managing the development of high-technology enterprises in both the private and public sectors. He received his B.S. and M.S. in aeronautics and astronautics and his Sc.D. in instrumentation from the Massachusetts Institute of Technology (MIT). Dr. Fraser joined MIT's Instrumentation Laboratory (which became the Charles Stark Draper Laboratory in 1973) as a member of the technical staff working on Apollo flight controls; later he served as the director of the Control and Flight Dynamics Division, vice president of technical operations, and executive vice president. From 1990 to 1991, Dr. Fraser was the deputy director of operational testing and evaluation for command, control, communications, and intelligence at the U.S. Department of Defense. After Senate confirmation he was appointed Principal Deputy Undersecretary of Defense (Acquisition) from 1991 to 1993. From 1993 until he retired in 2006, Dr. Fraser was the director of the Boston University Photonics Center and a professor of engineering and physics. His honors include membership in the National Academy of Engineering and receipt of the Defense Distinguished Service Medal; he is also an Honorary Fellow of the American Institute of Aeronautics and Astronautics. Dr. Fraser has served on the NASA Advisory Council, was a former member of the Aeronautics and Space Engineering Board, and has served as chair of several National Research Council (NRC) study groups, as well as being a member of many other NRC study groups.

Charles E. Adolph is currently an independent consultant and has approximately 50 years' experience in testing and evaluation and acquisition management. He started his career with General Dynamics Convair as a flight test engineer at Edwards Air Force Base, California, in 1956. Following 3 years in the U.S. Air Force, he held a variety of engineering and systems acquisition, technical, and management positions with the Air Force, advancing to technical director, the senior civilian position at the Air Force Flight Test Center. From 1987 to 1994, he held several positions in the Office of the Secretary of Defense. For most of that period he was the director of Test and Evaluation, Acquisition, and Technology. He also served as the acting director of Operational Test and Evaluation and acting director of Defense Research and Engineering. He was a senior vice president for Science Applications International Corporation (SAIC) from 1994 to 2000 and served as the manager of the SAIC Testing and Evaluation group. Mr. Adolph received a B.S. degree in aeronautical engineering from St. Louis University, an M.S. in aeronautical and astronautical engineering from the University of Michigan, and an M.S. in systems management from the University of Southern California, and he was a Sloan Fellow at the Stanford University Graduate School of Business.

Brian A. Arnold is the vice president of Space Strategy for Raytheon Company's Space and Airborne Systems (SAS) business. In this role, he determines evolving customer needs in the defense, intelligence, and civil arenas, and develops strategies to meet them with space-qualified solutions. He also leads planning efforts for expanding core SAS space markets and technologies. Before assuming his current position, Mr. Arnold served as the vice president and general manager of Space Systems within Raytheon SAS. A retired U.S. Air Force lieutenant general, he has 35 years of experience in leading space superiority programs and exceptional space market knowledge and expertise. Prior to joining Raytheon in 2005, Mr. Arnold served as commander, Space and Missile Systems Center, Air Force Space Command, Los Angeles Air Force Base, the nation's center of excellence for military space acquisition. There, he managed the research, design, development, acquisition, and sustainment of space launch and command and control systems, missile systems, and satellite systems. Mr. Arnold was commissioned through Officer Training School at Lackland Air Force Base, Texas, in 1971, and spent the majority of his Air Force career in operations as a pilot in FB-111 and B-52 aircraft; he has served as a commander at the flight, squadron, wing, and subunified level of command. As the director of Space and Nuclear Deterrence for the Assistant Secretary of the Air Force for Acquisition, he was responsible for space and missile systems. Mr. Arnold received a bachelor's degree in education from California State University, Hayward, and a master's degree in administrative education from Pepperdine University, Los Angeles.

Francis J. Baker is a professor of management at Wright State University, where he also directs Wright State University's Master of Business Administration (M.B.A.) program in project management. Prior to coming to Wright State University, Dr. Baker spent more than two decades in his previous career, as a United States Air Force officer: He served as a transport navigator, Minuteman missile launch-crew commander, Strategic Air Command staff officer, and U.S. Air Force Academy professor. In 1986, he came to the B-2 Stealth Bomber program at Wright-Patterson Air Force Base, Ohio. From 1986 until his departure in 1991, Dr. Baker was, at various times, the B-2 production program manager, chief of program integration, and executive officer to the B-2 program director. Since his arrival at Wright State in 1991, Dr. Baker has led the development of the university's popular project management M.B.A. program, and he is a columnist and contributing editor for the Project Management Institute's *PM Network* magazine. He has also received numerous teaching awards, including recognition as the outstanding teacher for the College of Business and Administration for 1994-1995 and Wright State's Presidential Award for Excellence in Teaching for 1997-1998. Dr. Baker received a Ph.D. and M.A. in management from the Peter F. Drucker School of the Claremont Graduate University, an M.B.A. from the University of North Dakota and a B.B.A. from St. John Fisher College.

Thomas W. Blakely is the vice president of engineering for Lockheed Martin Aeronautics Company. His career at Lockheed over the past three decades has spanned all three operating sites and multiple programs. He currently leads 7,300 engineers, scientists, and technicians throughout the company. Mr. Blakely joined Lockheed-California after graduating from Texas A&M University with a degree in aerospace engineering in 1979. As a young engineer, he was involved with a variety of development programs related to the P-3 Orion and CP-140 Aurora aircraft. He worked with the Maritime Patrol Engineering office at Naval Air Systems Command and later became the engineering program manager for the P-3C Orion programs. In 1991, Mr. Blakely transferred to Marietta, Georgia, assuming responsibility for all of Lockheed's International Maritime Patrol Aircraft Engineering programs. In 1996, he was selected to lead the C-130J systems verification and flight test team. He was subsequently promoted to chief systems engineer and ultimately chief engineer for C-130 programs. In August 2000, Mr. Blakely transferred to Fort Worth, Texas, to take the position of deputy for engineering, in which he coordinated the consolidation of engineering operations, personnel, processes, and tools across the newly formed Lockheed Aeronautics Company, which combined the operations of Palmdale, California; Marietta, Georgia; and Fort Worth, Texas. Special assignments as the technical director for the KC-130J and technical director for the C-5M development were followed by his being named vice president of engineering for Lockheed Martin Aeronautics Company in 2003. From early 2004 to mid-2006, he concurrently served as the company's technical director on the F-35 Joint Strike Fighter program. In addition, Mr. Blakely carved out time for continued education and participation in civic and professional organizations. In 2008, he earned a master's degree in systems engineering from Southern Methodist University, where he is currently working on his doctorate. Mr. Blakely serves as a board member for the Arts Council of Fort Worth and Tarrant County and participates in national organizations, accepting speaking engagements for groups such as the National Research Council, the American Institute of Aeronautics and Astronautics, and the American Society of Mechanical Engineers.

Claude M. Bolton is the executive-in-residence at the Defense Acquisition University (DAU). Mr. Bolton's primary focus is assisting the DAU president achieve the

Congressional direction to recruit, retain, train, and educate the Department of Defense (DoD) acquisition workforce. In addition, Mr. Bolton is an independent management consultant specializing in DoD program management, providing his expertise to DoD organizations and the defense industry. Mr. Bolton has had more than 30 years of experience in the business of acquisition, logistics, and technology, and his duties and experiences include being a fighter pilot, a combat pilot, and a test pilot. He has been a program manager on three acquisition category ID programs; commandant of Defense Systems Management College; Air Force Materiel Command Inspector General; Program Executive Officer for all Air Force fighters and bombers; and Air Force Security Assistance Center commander. Forty-eight hours after retiring from the Air Force in the rank of Major General, he became the Assistant Secretary of the Army for Acquisition, Technology, and Logistics and served in that position a history-making 6 years, acquiring everything for Army soldiers before retiring in January 2008 and assuming his current position at DAU. Mr. Bolton received his USAF commission in 1969 through the University of Nebraska's Air Force Reserve Officers' Training Corps Program, from which he was a distinguished graduate. Mr. Bolton's education includes a bachelor's degree in electrical engineering from the University of Nebraska, a master's degree in management from Troy State University, and a master's degree in national security and strategic studies from the Naval War College. In July 2006, he was awarded a Doctor of Science (Honoris Causa) degree from Cranfield University in England. In May 2007, he was awarded an Honorary Doctor of Science degree from the University of Nebraska-Lincoln (UNL), his alma mater. Mr. Bolton recently became the inaugural chair of the University of Nebraska's Space and Telecom Law Program Advisory Board. The UNL is the only U.S. university offering a degree in this area of growing importance.

Thomas J. Burns co-founded and serves as the chief executive officer and chair of SET Corporation, a research and development (R&D) company specializing in the development and commercialization of "smart sensing" technologies. Prior to founding SET, he co-founded and served as chief operating officer of ObjectVideo, Inc., a venture-backed leader in smart video solutions for commercial and military security applications. Dr. Burns joined ObjectVideo from the Defense Advanced Research Projects Agency (DARPA), where he pioneered the development of model-based signal and image exploitation technologies, building on his experiences directing computer vision research as a United States Air Force officer at the Air Force Research Laboratory (AFRL). While assigned to the AFRL, he led the laboratory's premiere Automatic Target Recognition program, receiving AFRL's prestigious Peter R. Murray Program Manager of the Year Award. Dr. Burns is co-inventor of patents on video and radar technology and has published numerous refereed papers in areas as diverse as electro-optics and wavelet mathematics. He holds a Ph.D. in electrical engineering from the Air Force Institute of Technology. Dr. Burns is a current member of the Air Force Studies Board.

Llewellyn S. Dougherty is the vice president, Special Programs, for Raytheon Company. He has served in other areas of the company, including sensors and communications, radar systems, and reconnaissance systems. Prior to his career at Raytheon, he was technical assistant to the director of the DARPA. His areas of expertise include avionics, digital computers, software, systems engineering, and systems safety. Dr. Dougherty received a Ph.D. in digital systems engineering from the Air Force Institute of Technology, an M.S. in aeronautics and astronautics from the Massachusetts Institute of Technology, and a B.S. in astronautics and engineering sciences from the U.S. Air Force Academy.

Richard B.H. Lewis is the vice president of Net-Centric Integration and Demonstration and a member of Lockheed Martin's Corporate Engineering and Technology (CE&T) organization. Mr. Lewis is responsible for building a corporate-wide infrastructure for modeling and simulation and for determining the standards and techniques that will guide future net-centric simulation, net-enabled warfare, visualization, and human-in-the-loop experimentation. He supports the operational-level assessments of business area experiments and exercises and helps develop a better understanding of complex missions and product capabilities. He leads an initiative to define and develop advanced capabilities for modeling and simulation, analysis, and demonstrations to support the corporation and its customers in the conduct of mission analysis to define requirements and address customer challenges. He has responsibility for the Global Vision Program, Lockheed Martin's corporate-wide capability, which enables the real-time development and demonstration of advanced, integrated technology concepts and solutions (classified and unclassified), and leverages this infrastructure to develop a framework that enhances cross-corporate collaboration and provides access to Lockheed Martin models, simulations, and tools. In addition, Mr. Lewis serves as the executive agent for Directed Energy and the executive sponsor for the Operations Analysis Community of Practice. Mr. Lewis comes to Lockheed Martin following a successful 35-year career with the Department of Defense, where he held a number of senior-level command and leadership positions, including director, Joint Theater Air and Missile Defense Organization; Program Executive Officer for fighter and bomber programs; and Program Executive Officer for the F-22 program at the U.S. Air Force Headquarters in Washington, D.C. He holds a bachelor's degree in mathematics and computer science from Colorado State University and a master's degree in systems management from the University of Southern California. A graduate of the U.S. Army War College, he retired from the USAF as a Major General.

Ellen M. Lord is senior vice president and general manager of AAI Corporation, an operating unit of Textron Systems Corporation (TSC) and an indirect wholly owned subsidiary of Textron, Inc. Ms. Lord joined AAI in April 2008, before which she served as vice president of integration management for TSC in Wilmington, Massachusetts. In that position, she led the team responsible for managing the integration of AAI into the Textron family of businesses—a process that will serve as a playbook for future acquisitions across Textron. Prior to that, she was the vice president of intelligent battlefield systems at Textron Defense Systems, an operating unit of TSC, where she was responsible for a business line including unattended networked ground sensor and munitions systems. She also served as the vice president of strategy for TSC, in addition to holding other tactical and strategic business and operations positions. Earlier in her career, Ms. Lord had managed proprietary and patented plastics technology for Textron Automotive Technology Center in Dover, New Hampshire. During her tenure, she led teams that developed an innovative new family of engineering thermoplastics for automotive interiors with Dow Plastics, as well as commercialized Bright Trim™, a revolutionary coating used by the U.S. "Big Three" automakers that looks like chrome and behaves like plastic. Ms. Lord earned a master's degree in chemistry from the University of New Hampshire as well as a Bachelor of Science degree in chemistry from Connecticut College. She also is a Textron Six Sigma certified Black Belt, specializing in Design for Six Sigma. Ms. Lord serves on the board of directors of the Greater Baltimore Committee, which has a membership of more than 500 organizations dedicated to increasing the competitiveness of the Baltimore business region; and on the Maryland Business Roundtable for Education, a group that unites the business community in support of quality education.

Christopher E. Manuel is the corporate vice president for the command, control, communications, computers, and networks (C4N) business area of the Sierra Nevada Corporation (SNC) in Oakland, California. He is responsible for the development and successful execution of the business area business management plan (profit and loss/marketing/programs and internal research and development). Mr. Manuel's prior responsibilities for SNC in San Francisco had included the overall direction and management of programs through designated program directors and/or program managers, ensuring consistency with corporate strategy, consistency of process across programs and projects, and customer satisfaction with the products and services provided; development and oversight of successful execution of the business unit business management plan (program/bids and proposals/internal research and development and marketing); and supporting the business unit lead in the execution and management of Capture Management and Planning (CMAP). Mr. Manuel is also a U.S. Army Special Forces Chief Warrant Officer 3 with experience in various countries, including Kuwait, Rwanda, Bosnia, and Afghanistan; he also served in Operation Iraqi Freedom as a consultant. Mr. Manuel is currently serving with the U.S. Army Reserves as an information systems technician for the Western Information Operation Center in Dublin, California. Mr. Manuel received a Bachelor of Arts degree in history from Fayetteville State University in North Carolina and a Master of Science degree in defense analysis from the Naval Postgraduate School in Monterey, California.

Matt L. Mleziva is currently the president of Wildwood Strategic Concepts, a strategic consulting company in Westford, Massachusetts. Mr. Mleziva has led joint Office of the Secretary of Defense (OSD) teams that developed recommendations projected to save millions of dollars annually. He guided Air Force Networked Tactical Communications efforts into a single joint program with the Navy. Mr. Mleziva has a proven track record of achieving cost, schedule, and performance goals across organizations covering a wide range of information system technologies for a diverse customer base. He acquired space, air, and electronic systems for the Department of Defense, the U.S. government, and foreign nations. Mr. Mleziva has demonstrated the capability to utilize emerging information technology and promote commonality and interoperability in combat systems. He developed ultra-streamlined acquisition strategy in response to urgent Air Force operational needs. Mr. Mleziva is the recipient of several awards, including the Presidential Meritorious Executive Rank Award and the Air Force Outstanding Civilian Career Service Award. He holds a post master's degree in electrical engineering, an M.S. in electrical engineering, and a B.S. in electrical engineering from the Massachusetts Institute of Technology. Mr. Mleziva is a current member of the Air Force Studies Board.

Ronald E. Mutzelburg is retired Washington, D.C., director for the Boeing Company's Phantom Works and Advanced Systems, a position that he had assumed when he joined Boeing in September 2002. His organization managed the relationship with senior U.S. government technology and advanced systems customers in Washington, D.C., including DARPA; the Office of the Director, Defense Research and Engineering; the Office of Naval Research; and NASA (Aeronautics); as well as the Office of the Secretary of Defense, Joint Staff, and military service technology and long-range capability requirements offices. Prior to joining Boeing, Mr. Mutzelburg completed a 34-year government career in the Department of Defense. From August 1992 to July 2002, he served as the deputy director for Air Warfare in the Office of Strategic and Tactical Systems, Undersecretary of Defense for Acquisition, Technology, and Logistics. He was responsible for acquisition oversight for the B-1, B-2, C-17, F-22, F-18, Joint Strike Fighter, Joint Surveillance Target Attack Radar System, unmanned air vehicles, several proprietary programs, and numerous air-to-air and air-to-ground weapons programs. From 1989 to July 2002, he was the assistant program director for the B-2 at Aeronautical Systems Division (ASD), Air Force Systems Command. From 1985 to 1989, he was the director of Fighter Propulsion in the Propulsion Systems Program Office, ASD. From 1983 to 1985, he was the director of Logistics, Propulsion Systems Program Office, ASD. From 1968 to 1982, he held numerous managerial and project officer assignments in the Air Force Logistics Command. He functioned in various professional specialties, including as an operations research analyst, cost analyst, logistics management specialist, computer programmer/analyst, and industrial engineer. Mr. Mutzelburg has a B.S.I.E. from Wayne State University (1968) and an M.S. in industrial and systems engineering from Ohio State University (1974) and is a graduate of National War College (1983).

Richard L. Rumpf is the president of Rumpf Associates International, Inc. (RAI). He is also a recognized expert in military (especially U.S. Navy) research, advanced technologies, and defense acquisition policy and procedures. In 1990, Mr. Rumpf established Rumpf Associates International, Inc., and serves as its president and chief executive officer. He has provided technical, programmatic, management services, and due-diligence analysis to a long list of commercial and government clients. He is recognized as an authority on defense procurement, requirements, management, and technology and has provided advice and expert testimony to a few law firms. He has a proven ability to review complex, multifaceted problems requiring technical, administrative, and political understanding; to get to the heart of a problem and quickly assess and prioritize the possible solutions; to organize and manage technical and acquisition personnel; to structure issues crisply; and to communicate effectively with congressional members, staffs, Office of the Secretary of Defense (OSD) principals, international ministries of defense, and industry leaders. Mr. Rumpf currently supports the Navy's initiative to leverage U.S. Department of Energy (DOE) modeling and simulation advancements to identify an alternative to the Navy's Full Ship Shock Test. He recently served as a consultant to the OSD Joint Ground Robotics Enterprise to assist with the establishment of the Robotics Technology Consortium and to develop the Department of Defense Unmanned Systems Integrated Roadmap. In 2000, Mr. Rumpf was a member of the DP-20 Self Assessment Team tasked by the Assistant Deputy Administrator for Military Application and Stockpile Operations for the National Nuclear Security Administration to identify opportunities to increase complementary work for the DOE weapons complex in order to help sustain required mission-critical skills and expertise capabilities. Prior to founding RAI, Mr. Rumpf served as the Acting Assistant Secretary of the Navy for Research, Engineering, and Systems. He managed $9 billion worth of annual research, development, testing, and evaluation resources for the U.S. Navy and developed, planned, and approved major acquisition programs acting as acquisition executive. In that capacity, he was responsible for planning and directing the research, development, engineering, testing, and evaluation of future weapons, sensors, ships, aircraft, unmanned vehicles, and space systems for the U.S. Navy and Marine Corps. Mr. Rumpf holds an M.S. degree in aerospace engineering from the University of Colorado, a diploma in fluid dynamics from the Von Karman Institute in Belgium, and a B.S. degree in aeronautical engineering from the University of Colorado.

Appendix B

Meetings and Participating Organizations

MEETING 1
MARCH 30-31, 2010
THE KECK CENTER OF THE NATIONAL ACADEMIES
WASHINGTON, D.C.

Evaluation of USAF Preacquisition Technology Development
Dr. Steven Walker, Senior Executive Service, Deputy Assistant Secretary of the Air Force for Science, Technology, and Engineering, Washington, D.C., SAF/AQR

DoD Requirements 101
Colonel Wayne McGee, USAF, Chief, Integration Division, Headquarters, Washington, D.C., A5RP

Acquisition 101
Mr. Jeffery R. Shelton, Associate Deputy Assistant Secretary, Acquisition Integration, Washington, D.C., SAF/AQX

AoA Process Introduction
and
Analysis of Alternatives for Recapitalizing the U.S. Air Force KC-135 Aerial Refueling Tanker Fleet
Dr. Michael Kennedy, Senior Economist, RAND Corporation

National Research Council Pre-Milestone A and Early-Phase Systems Engineering Report
General Lester Lyles, USAF (retired), Independent Consultant, The Lyles Group, Virginia

Development Planning (DP)
Brigadier General Dwyer L. Dennis, USAF, Director, Intelligence and Requirements Directorate, Headquarters Air Force Materiel Command, Wright-Patterson Air Force Base, Ohio

DoD/Air Force Competitive Prototyping
Dr. Clarence Gooden, Deputy Chief, Engineering and Technical Management Division, SAF/AQRE

DoD/AF Acquisition Laws, Regulations, and Policies
Mr. Richard Fowler, Program Learning Director (Acquisition Law), Defense Acquisition University

MEETING 2
APRIL 21-22, 2010
THE KECK CENTER OF THE NATIONAL ACADEMIES
WASHINGTON, D.C.

PPBE Executive Training: Course for the National Research Council Committee on Air Force Pre Acquisition Technology Development
Colonel Thomas Thurston, USAF (retired), Program Manager, PPBE Processes and Training Programs, Science Applications International Corporation

GAO Observations of Preacquisition Technology Development Practices
Mr. Michael J. Sullivan, Director, Acquisition and Sourcing Management Team
Mr. Bruce H. Thomas, Assistant Director, Acquisition and Sourcing Management Team
Ms. Cheryl K. Andrew, Senior Analyst, Acquisition and Sourcing Management Team

Capabilities Integration Directorate
and
Cyber Acquisition Strategy
Dr. Charles Kelley, Director, Capability Integration, Electronic Systems Center, Hanscom Air Force Base, Massachusetts, ESC/XR

Congressional Panel Discussion
Mr. Peter Levine, General Counsel for the Senate Armed Services Committee, United States Senate
Ms. Betsy Schmid, Professional Staff Member, Senate Appropriations Committee, Subcommittee on Defense, United States Senate

Pre-Systems Acquisition
Lieutenant General Richard M. Scofield, USAF (retired), Independent Consultant

ASC/XR DP Discussion
Colonel Edward M. Stanhouse, USAF, Director, Requirements and Capabilities Integration, Aeronautical Systems Center, ASC/XR

SMC/XR Development Planning
Colonel Donald E. Wussler, Jr., USAF, Director, SMC Developmental Planning, Space and Missile Systems Center, Los Angeles Air Force Base, California, SMC/XR

System Engineering within Defense Research and Engineering
Mr. Stephen P. Welby, Director, Systems Engineering, Office of the Secretary of Defense

SITE VISIT
MAY 3, 2010
AIR FORCE MATERIEL COMMAND
WRIGHT-PATTERSON AIR FORCE BASE
DAYTON, OHIO

Sponsor Discussion of Study Scope and Expectations
Dr. Steven Walker, Senior Executive Service, Deputy Assistant Secretary of the Air Force for Science, Technology, and Engineering, Washington, D.C., SAF/AQR

MEETING 3
MAY 12-13, 2010
THE NATIONAL ACADEMY OF SCIENCES
WASHINGTON, D.C.

Skunk Works Technology Development: Approaches and Issues
Colonel Neil Kacena, USAF (retired), Vice President, Advanced Development Programs Deputy, Lockheed Martin Aeronautics Company

Rapid Acquisition: JIEDDO Best Practices
Mr. Mitchell A. Howell, Senior Executive Service, Deputy Director, Rapid Acquisition and Technology, JIEDDO

Preacquisition Technology Development
Ms. Dawn Meyerriecks, Deputy Director of National Intelligence for Acquisition and Technology

Technology Development: Best Practices
Mr. A. Thomas Young, Executive Vice President (retired), Lockheed Martin

Development Planning Initiative Within DoD
The Honorable Zachary J. Lemnios, Director of Defense Research and Engineering

Rapid Prototyping at Scaled Composites
Mr. Douglas B. Shane, President, Scaled Composites, LLC

AFRL Perspective on Improving Technology Development and Transition
Dr. Michael A. Kuliasha, Chief Technologist, Air Force Research Laboratory, Wright-Patterson Air Force Base, Ohio

SITE VISIT
MAY 14, 2010
OFFICE OF THE SECRETARY OF THE AIR FORCE FOR SCIENCE, TECHNOLOGY, AND ENGINEERING
WASHINGTON, D.C.

Air Force Systems Engineering
Colonel Shawn Shanley, USAF, Chief, Engineering and Technical Management Division, SAF/AQRE

SITE VISIT
JUNE 1, 2010
AIR FORCE MATERIEL COMMAND
WRIGHT-PATTERSON AIR FORCE BASE
DAYTON, OHIO

ASC Involvement in Applied Technology Councils (ATCs)
Mr. Gerald L. Freisthler, Senior Executive Service, Executive Director, Aeronautical Systems Center
Colonel Arthur F. Huber, USAF, Vice Commander, Aeronautical Systems Center

MEETING 4
JUNE 7-8, 2010
THE KECK CENTER OF THE NATIONAL ACADEMIES
WASHINGTON, D.C.

Dragon Eye—A Small UAV from a Paper Sketch to an Operational System in 30 Months
Dr. Francis J. Klemm, Superintendent, Tactical Electronic Warfare Division, Naval Research Laboratory

Approaches to Rapid Acquisition of Revolutionary and Evolutionary Space Systems Capabilities
Dr. John P. Schaub, Superintendent, Spacecraft Engineering Division, Naval Research Laboratory

Transitioning Technologies into Enduring Solutions
Mr. Daniel G. Wolfe, Chief Executive Officer, Universal Solutions International, Inc.

The Role of MISSLEs in the Acquisition Process
Dr. Raymond R. Buettner, Jr., Director, Field Experimentation, Naval Postgraduate School

Improving Technology Development in Cyber: Challenges and Ideas
Dr. Jon Goding, Principal Engineering Fellow, Raytheon

Information Technology Acquisition
Major General Paul F. Capasso, USAF, Director, Network Services, Office of Information Dominance, and Chief Information Officer

3M Innovation Story
Mr. Thomas Gehring, Program Manager, 3M Industrial and Transportation Business

Sierra Nevada Corporation: A Commercial Space Company Case Study: Modular, Rapid, Repeatable
Mr. Mark Sirangelo, Corporate Vice President, Sierra Nevada Corporation, Space Systems Group

Remarks to the National Research Council Air Force Preacquisition Technology Development Study Panel
Mr. Benjamin P. Riley, Principal Deputy, Rapid Fielding Directorate, Office of the Assistant Secretary of Defense for Research and Engineering

SITE VISIT
JUNE 11, 2010
HOUSE ARMED SERVICES COMMITTEE
WASHINGTON, D.C.

Congressional Views Toward Air Force Technology Development
Mr. Douglas Roach, Professional Staff Member
Mr. John Oppenheim (GAO Detailee)
Ms. Lynne Williams, Professional Staff Member
Mr. Timothy McClees, Professional Staff Member
Mr. John Sullivan, Professional Staff Member
Mr. William Ebbs, Professional Staff Member

SITE VISIT
JUNE 11, 2010
AIR FORCE RESEARCH LABORATORY
WRIGHT-PATTERSON AIR FORCE BASE
DAYTON, OHIO

Discussion of USAF Science and Technology (S&T) Vision and Related Issues
Major General Ellen Pawlikowski, Commander, Air Force Research Laboratory

SITE VISIT
JULY 6, 2010
OFFICE OF THE SECRETARY OF THE AIR FORCE FOR SCIENCE, TECHNOLOGY, AND ENGINEERING
WASHINGTON, D.C.

Views on Technology Development
Mr. Steven Munday, Policy and Program Analyst, SAF/AQRS
Mr. Jerry Lautenschlager, Deputy Chief, Strategic Planning Branch, SAF/AQRS

MEETING 5
JULY 7-8, 2010
THE KECK CENTER OF THE NATIONAL ACADEMIES
WASHINGTON, D.C.

Writing Meeting

SITE VISIT
JULY 15, 2010
AIR FORCE MATERIEL COMMAND
WRIGHT-PATTERSON AIR FORCE BASE
DAYTON, OHIO

Perspectives on 6.4 Technology Development Funds; USAF S&T Strategy; Current DoD Competitive Prototyping Policy; Leadership Development; and the Weapon Systems Acquisition Reform Act of 2009
General Donald Hoffman, Commander, Air Force Materiel Command
Mr. Gregory Garcia, Deputy Director, Intelligence and Requirements Directorate, Air Force Materiel Command
Lieutenant Colonel Bill Gideon, USAF, Executive Officer, AFMC/CCE

SITE VISIT
JULY 27, 2010
DEFENSE ADVANCED RESEARCH PROJECTS AGENCY
WASHINGTON, D.C.

Perspectives on Technology Development for Future Weapon Systems
Mr. Ellison Urban, Director, Adaptive Execution Office

SITE VISIT
JULY 29, 2010
OFFICE OF STUDIES AND ANALYSES, ASSESSMENTS, AND LESSONS LEARNED
HEADQUARTERS U.S. AIR FORCE
WASHINGTON, D.C.

Perspectives on Prioritization and Allocation of Air Force Research Funding and the Tie Between Research and Development and Future Force Structure Requirements
Dr. Jacqueline Henningsen, Director, Office of Studies and Analyses, Assessments, and Lessons Learned

COMMITTEE CONFERENCE CALL
AUGUST 10, 2010

Debriefings of Recent Committee Site Visits

MEETING 6
AUGUST 24-26, 2010
J. ERIK JONSSON CENTER
WOODS HOLE, MASSACHUSETTS

Writing Meeting

Appendix C

Background Information on Policies and Processes Related to Technology Development

PLANNING, PROGRAMMING, BUDGETING, AND EXECUTION SYSTEM

The Planning Phase

The planning phase of the Planning, Programming, Budgeting, and Execution System (PPBES) begins with the Office of the Secretary of Defense (OSD) and the Joint Staff collaboratively articulating resource-informed national defense policies and military strategy known as the Strategic Planning Guidance (SPG).[1] The SPG then shapes the "Enhanced" Planning Process (EPP). The result of EPP is a set of budget-conscious priorities for program development (military force modernization, readiness, and sustainability; and supporting business processes and infrastructure), and is written up in the Joint Programming Guidance (JPG). Figure C-1 links the planning and programming phases and helps Department of Defense (DoD) departments and agencies write their Program Objective Memorandums (POMs).

The Programming Phase

For the United States Air Force (USAF), the programming phase begins with the writing of the POM. The POM balances program budgets as set down in the

[1] DAU Web site. Available at https://dap.dau.mil/aphome/ppbe/Pages/Default.aspx. Accessed August 10, 2010.

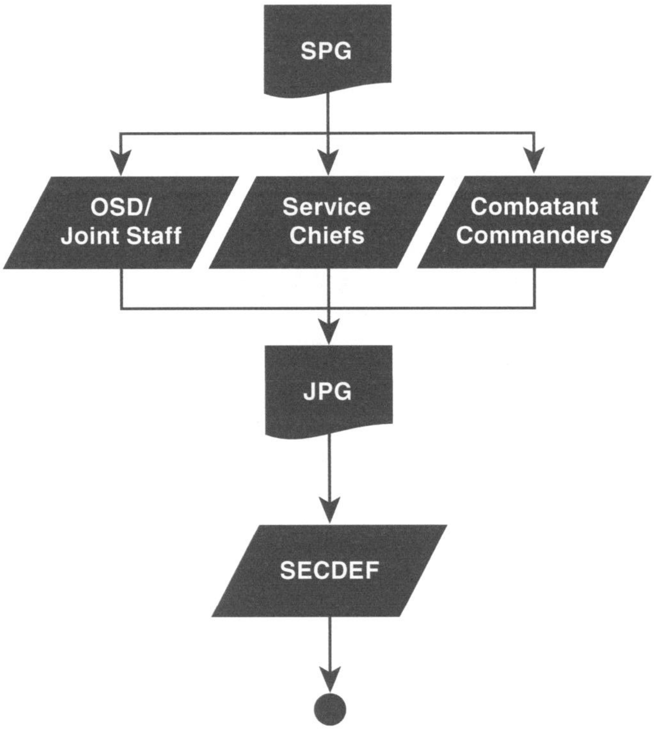

FIGURE C-1
The planning and programming phases' linkage to Department of Defense departments. SOURCE: DAU. 2010. *The Planning PPBE Phase.* Available at https://dap.dau.mil/aphome/ppbe/Pages/Planning.aspx. Accessed August 10, 2010.

JPG. When complete, the POM describes, in detail, the proposed budget (forces, personnel, and funding) for the next 6 years.[2] The POM may also describe what is not fully funded and the risks associated with the budget shortfall. Senior leaders of the OSD and Joint Staff review all of the military service POMs and integrate them into an overall coherent DoD program. For any shortfalls or any other issues with any portion of any POM, the OSD and Joint Staff can propose alternatives and marginally adjust budgets. As shown in Figure C-2, the Secretary of Defense settles any unresolved issues and writes them up in a Program Decision Memorandum (PDM).

The Budgeting Phase

The budgeting phase of the PPBES happens at the same time as the programming phase. Each DoD department and agency submits its budget estimate with its POM, and converts its program budget into the congressional appropriation structure format and submits it, along with justification. The budget forecasts

[2]DAU. 2010. *The Planning PPBE Phase.* Available at https://dap.dau.mil/aphome/ppbe/Pages/Planning.aspx. Accessed August 10, 2010.

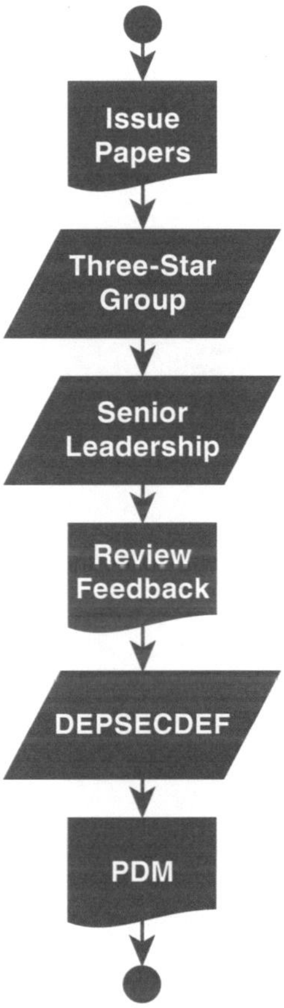

FIGURE C-2
The programming phase of Planning, Programming, Budgeting, and Execution (PPBE). SOURCE: DAU. 2010. *The Programming PPBE Phase.* Available at https://dap.dau.mil/aphome/ppbe/Pages/Programming.aspx. Accessed August 10, 2010.

only the next 2 years, but with more detail than in the POM. Office of the Under Secretary of Defense (Comptroller) and Office of Management and Budget (OMB) analysts review the submissions to ensure that program funding matches current policy, that individual programs are priced correctly, and that each program is fully justified to Congress. Typically, the analysts write up their questions during formal budget review hearings. After the hearings, each analyst prepares a Program Budget Decision (PBD) for each program, which proposes financial adjustments to address any issues or problems identified during the hearing. The PBD then goes to the Deputy Secretary of Defense for a decision. The decision goes in an updated budget submission to the OMB (as shown in Figure C-3). After that, the overall DoD budget is provided as part of the President's Budget Request to Congress.[3]

Thus the PPBE process addresses preacquisition technology development only indirectly, and it delegates the responsibility for the review of individual program

[3]Ibid.

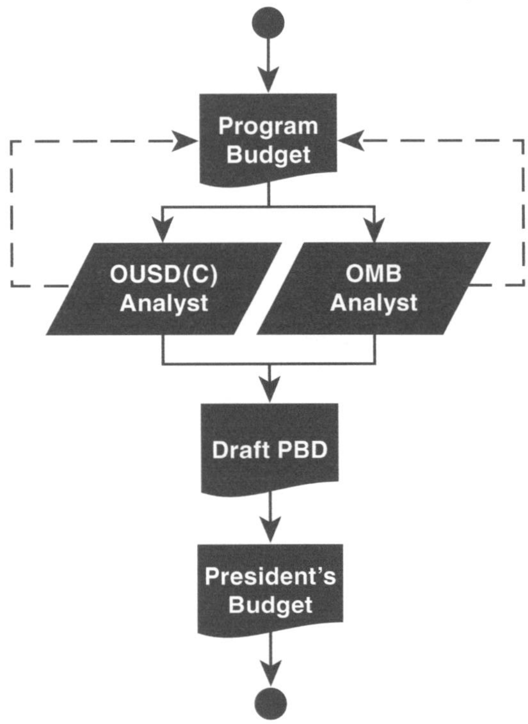

FIGURE C-3
The budgeting phase of Planning, Programming, Budgeting, and Execution (PPBE). SOURCE: DAU. 2010. *The Budgeting PPBE Phase.* Available at https://dap.dau.mil/aphome/ppbe/Pages/Budgeting.aspx. Accessed August 10, 2010.

activities to the services as part of their POM development, with OSD-level review only when adjustments are addressed during formal budget review hearings. The overall prioritization of USAF preacquisition technology development activities is handled during POM formulation and in detail only by the Air Staff Board Panels, and within each panel only for those Program Elements that are assigned.

DEPARTMENT OF DEFENSE INSTRUCTION 5000.02

The policy document DoD Instruction 5000.02 was released in December 2008. It updated the 5000 series policy document in order to incorporate congressionally mandated acquisition changes contained in the Weapon Systems Acquisition Reform Act (WSARA; Public Law 111-23), which was signed into law in May 2009.[4] It also incorporated a number of policy memorandums that had previously been issued by the Undersecretary of Defense for Acquisition, Technology, and Logistics. One example is a policy memorandum on competition and prototyping.[5]

[4]Available at http://www.ndia.org/Advocacy/PolicyPublicationsResources/Documents/WSARA-Public-Law-111-23.pdf. Accessed August 10, 2010.

[5]DoD. 2007. "Memorandum for Secretaries of the Military Departments, Chairman of the Joint Chiefs of Staff, Commander, U.S. Special Operations Command, Directors of the Defense Agen-

Although DoD Instruction 5000.02 discusses the preacquisition phase, it provides little "how-to" guidance, nor has it generated any formal direction either to train the workforce or to employ DoD Instruction 5000.02 effectively and measure the resulting ability of the workforce to execute acquisition programs successfully.

Two significant changes highlighted in DoD Instruction 5000.02 are (1) the requirement for competitive prototyping on all Major Defense Acquisition Programs (MDAPs), and (2) the relocation of the Preliminary Design Review (PDR) from taking place after Milestone B to occurring before Milestone B. This leads to a consequence (perhaps unintended) that a significant increase in research and development (R&D) investment is now required to fund the competitive prototyping efforts, as well as the necessary increased technical effort to define the product at a level of detail sufficient to meet PDR requirements. The policy does allow competitive prototyping to be waived, but, as discussed in Chapter 2 and Chapter 4, the process to obtain a waiver is not defined. Also, the timing of the PDR can be changed if approved by the Milestone Decision Authority.

AIR FORCE ACQUISITION IMPROVEMENT PLAN

As stated in Chapter 2 of this report, the USAF Acquisition Improvement Plan (AIP) does not directly address technology development for any phase of the acquisition life cycle. The purpose of the AIP, which was signed out to the field by the Secretary of the Air Force and the Air Force Chief of Staff, is explained in Box C-1.

The five initiatives that constitute the AIP[6] and on which the Air Force has focused its attention and implementation activities to date include:

1. Revitalize the Air Force acquisition workforce.
2. Improve requirements generation process.
3. Instill budget and financial discipline.
4. Improve Air Force major source selections.
5. Establish clear lines of authority and accountability within acquisition organizations.[7]

---

cies. Subject: Prototyping and Competition." Memorandum from the Under Secretary of Defense, September 19, 2007. Washington, D.C.: DoD. Available at https://dap.dau.mil/policy/Documents/Policy/20070921%20Prototyping%20and%20Competition%20ATL.pdf. Accessed August 10, 2010.

[6]USAF. 2009. *Acquisition Improvement Plan.* Washington, D.C.: Office of the Assistant Secretary of the Air Force (Acquisition). Available at http://www.dodbuzz.com/wp-content/uploads/2009/05/acquisition-improvement-plan-4-may-09.pdf. Accessed August 10, 2010.

[7]Extracted from "Air Force Officials Unveil New Acquisition Plan," Tech. Sgt. Amaani Lyle. Secretary of the Air Force Public Affairs. May 11, 2009. Available at http://www.af.mil/news/story.asp?id=123148399. Accessed on January 12, 2011.

BOX C-1
The Purpose of the Acquisition Improvement Plan

MEMORANDUM FOR ALMAJCOM-FOA-DRU/CC
DISTRIBUTION C
MAY 4, 2009

SUBJECT: Air Force Acquisition Improvement Plan

"The United States Air Force is committed to recapturing acquisition excellence by rebuilding an Air Force acquisition culture that delivers products and services as promised-on time, within budget and in compliance with all laws, policies and regulations. To do so, we have developed the attached Acquisition Improvement Plan."

"Our Challenge: Recapturing acquisition excellence requires an experienced, skilled, empowered, and accountable workforce, and begins with proper requirements and adequate and stable funding. The following five initiatives and their associated actions set forth a comprehensive improvement plan for addressing the foregoing acquisition issues."

SOURCE: Extracted from USAF 2009. "Acquisition Improvement Plan." Washington, D.C.: Air Force. Available at http://www.dodbuzz.com/wp-content/uploads/2009/05/acquisition-improvement-plan-4-may-09.pdf. Accessed August 10, 2010.

A close review of the five initiatives shows little focus on the preacquisition technology development phase.

JOINT CAPABILITIES INTEGRATION AND DEVELOPMENT SYSTEM

The Joint Capabilities Integration and Development System (JCIDS) plays a key role in identifying the capabilities required by the warfighters to support the National Defense Strategy, the National Military Strategy, and the National Strategy for Homeland Defense. The successful delivery of those capabilities relies on the JCIDS process working in concert with other joint and DoD decision processes. The procedures established in JCIDS support the Chairman of the Joint Chiefs of Staff and the Joint Requirements Oversight Council (JROC) in advising the Secretary of Defense in identifying and assessing joint military capability needs. The DoD has adopted Joint Capability Areas (JCAs) as its capability management language and framework. JCAs are collections of like DoD capabilities, functionally grouped to support capability analysis, strategy development, investment decision making, capability portfolio management, and capabilities-based force development and operational planning. JCIDS uses the JCA as an organizing construct. The Functional Capabilities Boards are organized around the Tier 1 JCA, and the JCIDS documents link the capabilities identified to the applicable JCA.

Introduction to the JCIDS Process

A depiction of the relationship between the JCIDS process and key acquisition decision points is provided in the Figure C-4. The JCIDS process is closely linked to the Defense Acquisition System.[8]

The JCIDS process was created to support the statutory responsibility of the JROC to validate joint warfighting requirements. The JCIDS is also a key supporting process for DoD acquisition and Planning, Programming, Budgeting, and Execution (PPBE) processes. The primary objective of the JCIDS process is to ensure that the capabilities required by the joint warfighter are identified with their associated operational performance criteria so that the assigned missions can be successfully executed. This is done through an open process that provides the JROC with the information that it needs in order to make decisions on required capabilities. The JCIDS process supports the acquisition process by identifying and assessing capability needs and associated performance criteria to be used as a basis for acquiring the right capabilities, including the right systems. These capability needs then serve as the basis for the development and production of systems to fill those needs. Additionally, the JCIDS provides the PPBE process with affordability advice by assessing the development and production life-cycle cost.

During the technology development phase, the sponsor performs technology maturation activities, builds competitive prototypes, and may perform design activities leading to a Preliminary Design Review. The Initial Capabilities Document provides a wide aperture for operational capability to define system requirements and to encourage technological innovation.[9] It is vital that the science and technology, users, training, and system developer communities collaborate to agree on a proposed solution that is affordable, militarily useful, and based on mature, demonstrated technology.

---

[8]CJCS. 2009. *Joint Capabilities Integration and Development System. Chairman of the Joint Chiefs of Staff Instruction.* CJCSI 3170.01G. March 1, 2009. Washington, D.C.: JCS. Available at http://www.dtic.mil/cjcs_directives/cdata/unlimit/3170_01.pdf. Accessed August 10, 2010.

[9]For a definition of "Initial Capabilities Document," see Part II of Enclosure D in: CJCS. 2009. *Joint Capabilities Integration and Development System. Chairman of the Joint Chiefs of Staff Instruction.* CJCSI 3170.01G. March 1, 2009. Washington, D.C.: JCS. Available at http://www.dtic.mil/cjcs_directives/cdata/unlimit/3170_01.pdf. Accessed August 10, 2010.

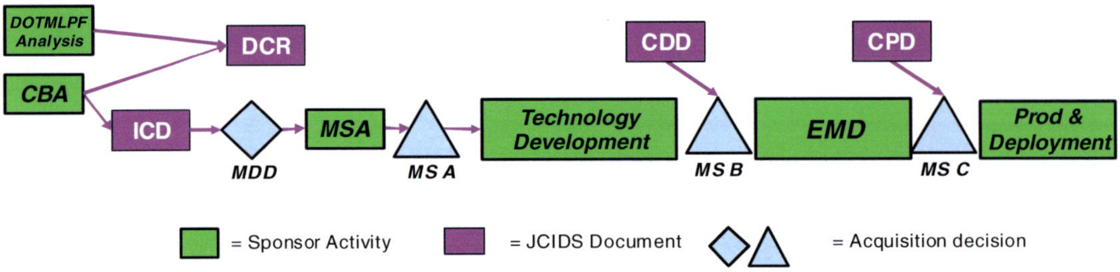

FIGURE C-4
Joint Capabilities Integration and Development System (JCIDS) process and key acquisition decision points. SOURCE: Joint Chiefs of Staff (JCS). 2009. *Joint Capabilities Integration and Development System. Chairman of the Joint Chiefs of Staff Instruction.* CJCSI 3170.01G. March 1, 2009. Washington, D.C.: JCS. Available at http://www.dtic.mil/cjcs_directives/cdata/unlimit/3170_01.pdf. Accessed August 10, 2010.

DEPARTMENT OF DEFENSE AND AIR FORCE COMPETITIVE PROTOTYPING

The Air Force has long used competitive prototyping with varying degrees of success, notably for aircraft and weapons development. For example, in the 1970s, the Air Force successfully managed the design and flight test of six new designs in three competitive prototype programs; that is, the YF-16/17, YA-9/10, and YC-14/15 aircraft. In the past several years, competitive prototyping has been effectively employed in a variety of air-to-surface weapons development programs, including the Joint Direct Attack Munition and Small Diameter Bomb.[10] Key factors in the effective use of prototyping, either competitive or noncompetitive, include a definition and disciplined assessment of the critical technologies to be demonstrated, in addition to the specification of a few key system performance parameters. A structured test and evaluation program is critical in order to demonstrate each prototype's capability and provide data for a comprehensive and objective assessment of the maturity of critical technologies. Additionally, discipline must be imposed downstream of the prototyping effort to minimize requirements changes.

In the past 3 years, there has been congressional legislation and OSD direction reinforcing the requirement for competitive prototyping when appropriate. In a September 19, 2007, memorandum, the Under Secretary of Defense for Acquisition, Technology, and Logistics directed that "Military Services and Defense Agencies will formulate all pending and future programs with acquisition strategies and fund-

[10]GAO. 2010. *Strong Leadership Is Key to Planning and Executing Stable Weapon Programs.* GAO-10-522. Washington, D.C.: GAO. Available at http://www.ndia.org/Advocacy/LegislativeandFederalIssuesUpdate/Documents/May2010/GAO-StrongLeadershipinWeaponssystemsprograms.pdf. Accessed August 10, 2010.

ing that provide for two or more competing teams producing prototypes through Milestone (MS) B."[11] WSARA requires competitive prototyping of systems before Milestone B, unless waived by the Milestone Decision Authority. The legislation further requires that a prototype be produced before Milestone B even if *competitive* prototyping is waived. The WSARA direction on prototyping has been incorporated in the *Defense Acquisition Guidebook* (DAG)[12] as well as in DoD Instruction 5000.02.[13] The DAG requires that the technology development strategy include a description of the prototyping strategy at the system and subsystem levels, as well as the number of prototype units that may be produced and employed during technology development and competitive prototyping.

The Air Force has responded to the congressional and OSD direction by updating internal guidance on competitive prototyping including AFI 63-101 on Prototyping.[14] The Air Force prototype vision recognizes that risk is not limited to technology, but that it includes integration and manufacturability risks as well. The policy specifies that prototypes should be considered for critical technology elements, key manufacturing and integration risks, and demonstration of the ability of the planned system to meet user requirements.

[11]DoD. 2007. "Memorandum for Secretaries of the Military Departments, Chairman of the Joint Chiefs of Staff, Commander, U.S. Special Operations Command, Directors of the Defense Agencies. Subject: Prototyping and Competition." Memorandum from the Under Secretary of Defense, September 19, 2007. Washington, D.C.: DoD. Available at https://dap.dau.mil/policy/Documents/Policy/20070921%20Prototyping%20and%20Competition%20ATL.pdf. Accessed August 10, 2010.

[12]DAU. 2010. *Defense Acquisition Guidebook.* Fort Belvoir, Va.: Defense Acquisition University. Available at http://www.ndia.org/Advocacy/LegislativeandFederalIssuesUpdate/Documents/March2010/Defense_Acqauisition_Guidebook_3-10.pdf. Accessed August 11, 2010.

[13]DoD. 2008. *Department of Defense Instruction. Subject: Operation of the Defense Acquisition System.* 5000.02. Washington, D.C.: Office of the Under Secretary of Defense for Acquisition, Technology, and Logistics. Available at http://www.dtic.mil/whs/directives/corres/pdf/500002p.pdf. Accessed August 11, 2010.

[14]USAF. 2010. *Air Force Guidance Memorandum to AFI 63-101: Acquisition and Sustainment Life Cycle Management.* Washington, D.C.: Department of Defense. Available at http://www.af.mil/shared/media/epubs/AFI63-101.pdf. Accessed August 11, 2010.

Appendix D

Background Information on the Vanguard Process and Applied Technology Councils

There is little, if anything, in the current acquisition environment relating to technology development that has not arisen—and subsequently been addressed—many times before. But organizational memory can be a fleeting thing, and the lessons learned today may be forgotten tomorrow. Shifting requirements, inadequate processes, chronic funding issues, and excessive oversight have existed as long as there has been a United States Air Force (USAF)—and even before. In his 1949 autobiography, General of the Air Force H.H. "Hap" Arnold complained:

> The tough part of aircraft development and securing an air program is to make Congress, the War Department, and the public realize that it is impossible to get a program that means anything unless it covers a period of not less than five years. Any program covering a shorter period is of little value. Normally it takes five years from the time the designer has an idea until the plane is delivered to the combatants. The funds must cover the entire period or there is no continuity of development or procurement. For years, the Army—and the Army Air Forces while a part of it—was hamstrung in its procurement programs by governmental shortsightedness.[1]

Around the time that General Arnold was writing those words, the newly established Air Force was creating the Ridenour Committee to study the USAF's research and development activities. The Ridenour Committee recommended the creation of a new organization, separate from the Air Materiel Command, to control all of the USAF's research and development. By the mid-1950s, there was recognition that

[1] Henry Harley "Hap" Arnold. *Global Mission.* New York: Harper, p. 156.

formal channels were needed to connect Combatant Commands, the science and technology (S&T) community, and the Product Centers. These ideas resulted in the establishment in 1960 of an organization called the Advanced System Program Office, developing mission requirement analysis and operational assessment tools and using them to focus technology development.[2]

THE DEVELOPMENT OF VANGUARD

The first Development Planning (DP) offices were begun in the 1960s, and their processes and policies were defined over the following years. In 1978, the Commander of Air Force Systems Command, General Alton D. Slay, created Vanguard, a more comprehensive and complex DP methodology. With Vanguard, General Slay split the management of technology into two pieces. The first, what he called "Planning for Development," was *acquisition-based*: It codified user requirements and determined the systems, costs, schedules, and plans necessary to meet those requirements.

The second piece was called "Development Planning," and it was *technology based*, coordinating *all* research and development in the Air Force, focusing on Exploratory Development (6.2) and Advanced Development (6.3). Vanguard used what must have been for the time very advanced computer tools to increase visibility into technology efforts across all fronts, throughout industry and across the armed services. A channel was established within the Air Force Systems Command (AFSC), from the Deputy Chief of Staff for Development Plans down to each individual program office and laboratory, through which Vanguard data were accumulated, sorted, analyzed, and redistributed. Participation was not optional.

Essential to Vanguard's success was a tool called "Hooks and Strings," which formed the connective tissue between the Combatant Commands, the S&T world, and the acquisition centers. In connecting the three worlds, "Hooks and Strings" provided the answers to the critical questions that are as relevant today as they were 30 years ago. See Box D-1.

Project Vanguard included three core planning areas: (1) mission plans, (2) major force elements, and (3) functional plans. Mission-level plans addressed specific tasks that must be completed, whereas major force elements included larger and more general categories of systems that would garner interest across the board, and functional plans addressed those activities that spanned several mission areas. All of these plans included a wealth of information, such as applicable citations from the USAFs out-year development plan, relevant regulations, pertinent organizational dependencies, and proposed milestones and requirements.

[2]John M. Griffin and James J. Mattice. 2010. "Development Planning and Capability Planning, 1947 to 1999 and Beyond." Unpublished manuscript, pp. 5-6.

BOX D-1
The Objectives of Vanguard

1. The Vanguard "Hooks & Strings" Tool Provided Answers to the Following Questions:
   a. Do all USAF advanced development (6.3) projects have a clear and recognized trace back to some stated USAF capability deficiency or operational requirement?
   b. Do all advanced development (6.3) projects have a clear and recognized trace forward to some on-going, planned or-projected engineering/manufacturing·development program or project?
   c. Do all advanced development (6.3) project funding profiles and schedules take into account the schedules of EMD programs/projects which they support?
   d. Do all USAF *exploratory* development (6.2) projects have a clear trace to some existing/projected and officially recognized technology shortfall?
   e. Do all USAF *exploratory* development (6.2) projects have a clear and officially recognized "path" to advanced development or to some other exploitation of the generated technology?
   f. Do all defense industry IRAD projects supported directly by USAF funds have a clear trace directly to some existing and officially recognized technology shortfall which, if filled, would enhance the ability of the USAF to perform its mission?
   g. Can assurance be provided that technology work accomplished or underway by the USAF laboratories is not duplicated in contracts issued to defense contractors by USAF program offices?
   h. Can assurance be provided that each USAF EMD Program Office fully recognizes and exploits the technology accomplishments and advances made by the USAF laboratories which are applicable to the EMD program/project?
   i. Can USAF EMD Program Office be provided access to "entry level" information from all sources (USAF labs, other USAF programs/projects, other Services, Defense Industry) which *identifies* all available technology specifically related to their program/project?

SOURCE: Derived from an undated talking paper by Gen Alton D. Slay (USAF, Ret.) entitled "Vanguard."

Key parts of Vanguard were frequent, regular, face-to-face meetings at the four-star level, to facilitate coordination among all parties. In the words of General Slay:

> I hosted separate meetings each quarter at HQ AFSC with the operational commanders (e.g., SAC, MAC, TAC) and selected members of their staffs. Vanguard briefings described the Vanguard "hooks and strings" trace to all projects/ programs underway or planned in response to their requirements. Project funding levels and schedules were discussed in detail and comments solicited thereon.[3]

[3]Ibid., p. 4.

The results of Vanguard were mixed, according to its creator. Asked about his level of satisfaction with the Vanguard implementation as of his retirement in 1981, General Slay said:

> On the whole, I would rate my degree of satisfaction with its implementation as something just north of lukewarm. Maybe if I had been able to stick with it another year. . . .[4]

The loss of momentum described by General Slay is an important feature of this story. From Hap Arnold to General Slay to today's USAF leadership, a common thread emerges again and again: A leader sees the need for a better system to integrate the warfighter-S&T-acquisition worlds and creates a new management system to fill that need; the new system is developed and implemented, but soon the sponsoring leader moves on; a new commander arrives, together with that person's own views and priorities, and the departed leader's creation is treated with something like benign neglect; after a while, the old system is quietly dropped, and its lessons are soon forgotten.

Such was the case with Vanguard. Its demise is neither well documented nor well remembered, but it certainly did not survive the end of Air Force Systems Command in 1992, when AFSC essentially was subsumed within the old Air Force Logistics Command, and the new Air Force Materiel Command emerged.

APPLIED TECHNOLOGY COUNCILS

When it became clear that the need for coordination between the worlds of warfighter, S&T, and acquisition commands still existed, a new concept evolved at the Product Center level. Applied Technology Councils (ATCs) were instituted by Product Center and laboratory commanders to carry on the old Vanguard mission of integrating warfighter requirements with acquisition priorities and laboratory efforts. As with the Vanguard meetings, ATCs were held quarterly, attended by senior-level warfighters, top laboratory management, and high-level acquisition leaders. Warfighters made clear their combat requirements, S&T leaders explained what was feasible technologically, and the acquisition community set forth programmatic plans for matching requirements with new systems or subsystems. Priorities were established, funding was committed, and plans were made to transition technologies from the S&T world, over the "Valley of Death" to operational success—all as in the days of General Slay's Vanguard.

As with Vanguard, however, the ATCs have been allowed to erode past the point of usefulness, at least in some instances. The causes were many: Different commands had different assessments of the value of the ATC process. New com-

[4]Ibid., p. 6.

manders—whether warfighters, laboratory leaders, or acquisition top management—sometimes had other priorities. Sometimes overtaxed leadership let the intervals between ATCs increase, from quarterly to semiannual, then to annual, and sometimes beyond that. The staffs of participating organizations began to require multiple pre-briefings, adding bureaucracy to the process and arguably watering down the frank dialogue. Eventually, the rank—and the perspective and the power—of ATC attendees declined: What had at one time been meetings of lieutenant generals eventually became meetings of lieutenant colonels.

Vanguard and ATCs were both strong efforts aimed at integrating the needs and capabilities of operational commands (or Major Commands), S&T organizations, and acquisition centers. Both enjoyed success, and both eventually faded from view. As with the demise of Vanguard—and as with the declines of systems engineering and Development Planning—the erosion of ATCs in some areas represents a significant setback in the pursuit of a fully integrated technology development and systems acquisition mission.